T0235256

Lecture Notes in Artificial Intelligence 9717

Subseries of Lecture Notes in Computer Science

More information about this series at http://www.springer.com/series/1244

Ollivier Haemmerlé · Gem Stapleton
Catherine Faron Zucker (Eds.)

Graph-Based Representation and Reasoning

22nd International Conference
on Conceptual Structures, ICCS 2016
Annecy, France, July 5–7, 2016
Proceedings

 Springer

Editors
Ollivier Haemmerlé
Université Toulouse
Toulouse
France

Catherine Faron Zucker
Université Nice Sophia Antipolis
Nice
France

Gem Stapleton
University of Brighton
Brighton
UK

ISSN 0302-9743   ISSN 1611-3349 (electronic)
Lecture Notes in Artificial Intelligence
ISBN 978-3-319-40984-9   ISBN 978-3-319-40985-6 (eBook)
DOI 10.1007/978-3-319-40985-6

Library of Congress Control Number: 2016942019

LNCS Sublibrary: SL7 – Artificial Intelligence

Printed on acid-free paper

This Springer imprint is published by Springer Nature
The registered company is Springer International Publishing AG Switzerland

Welcome from the Conference Chairs

It was our pleasure to welcome you to the 22nd International Conference on Conceptual Structures, which was held at the University of Savoie in Annecy, France. The conference location offered ample opportunities for sightseeing and outdoor activities with the Mont Blanc summit nearby.

The idea of a colocation of ICCS and the International Conference on Formal Ontology in Information Systems (FOIS) in the French Alps germinated in Mumbai in January 2013, during ICCS 2013, whose audience was quite small. We wanted to revitalize ICCS by renewing the Program Committee around a preserved historic core, in order to build relationships with new communities. ICCS 2014 in Iasi in Romania was an important first step in that direction. For ICCS 2016, we assembled an excellent technical program, and the colocation with the 9th International Conference on Formal Ontology in Information Systems offered even more opportunities to attend high-quality research presentations and additional workshops. The co-location, which took place in Annecy, was unanimously supported by the ICCS Steering Committee, and is a beautiful concretization of our will.

Two highlights of this year's program were the keynote addresses delivered by two renowned researchers. Dr Mateja Jamnik, from the Cambridge University Computer Laboratory, has made significant research contributions to the diagrams and automated reasoning communities. Her research aims to investigate and mechanize human mathematical reasoning, such as the use of diagrams in proofs. Dr. Jamnik delivered a keynote talk entitled "Automating Human-Like Reasoning with Diagrams on Machines." The topic of her talk reflected our drive to broaden the community from which ICCS aims to attract both paper submissions and delegates attending the conference. Dr. Fabien Gandon is Research Director in Informatics and Computer Science at Inria Sophia Antipolis méditerranée and the Inria representative at the World-Wide Web Consortium (W3C). His professional interests include: Web, Semantic Web, social Web, ontologies, knowledge engineering and modelling, mobility, privacy, context-awareness, semantic social network/semantic analysis of social network, intraweb, distributed artificial intelligence. Dr Gandon delivered a keynote entitled "*On the Many Graphs of the Web and the Interest of Adding Their Missing Links.*"

ICCS 2016 received 40 full-paper submissions from authors spanning 23 countries. These papers were independently and anonymously reviewed, typically by four members of the Program Committee. After the reviews and an initial round of discussions among the reviewers and program chairs, the authors were invited to submit rebuttals. To ensure a high-quality program, the rebuttals were discussed and final versions of the reviews were produced. At the end of this process, the program chairs selected 14 full papers, giving a competitive acceptance rate of 35 %. A further five short papers were accepted for presentation and publication alongside six posters.

As with any organizational task of this scale, very many people contributed to the success of ICCS 2016. We would especially like to thank the members of the Program

Committee for their expert reviews and for their excellent engagement with the two discussion phases. The Program Committee also invited other experts to provide reviews for some papers, and we are grateful for their important contributions. We would also like to thank the following individuals: Richard Dapoigny and Patrick Barlatier, who made all of the local arrangements; Juliette Dibie, Liliana Ibanescu, Stéphane Dervaux for liaising with Springer and compiling the proceedings; Cassia Trojahn dos Santos for taking care of the conference finances and sponsorship; Sven Linker for publicizing the conference; and Nicolas Seydoux and Rémi Cavallo for being the webmaster and website designer, respectively. We are also grateful to the Steering Committee for their support and for entrusting us to organize ICCS 2016: Madalina Croitoru, Frithjof Dau, Ollivier Haemmerlé, Uta Priss, and Sebastian Rudolph. Finally, we thank our sponsors: Université Savoie Mont Blanc and Inria for financially supporting the conference organization, and AFIA, the French Society on AI, for sponsoring the Best Paper Award.

May 2016   Ollivier Haemmerlé
Catherine Faron Zucker
Gem Stapleton

Organization

ICCS Executive Committee

General Chair

Ollivier Haemmerlé   Université Toulouse Jean Jaurès, France

Program Co-chairs

Gem Stapleton   University of Brighton, UK
Catherine Faron-Zucker   Université Nice Sophia Antipolis, France

Local Chairs

Richard Dapoigny   Université de Savoie, France
Patrick Barlatier   Université de Savoie, France

Steering Committee

Madalina Croitoru   Université Montpellier 2, France
Frithjof Dau   SAP Research Dresden, Germany
Ollivier Haemmerlé   Université Toulouse Jean Jaurès, France
Uta Priss   Ostfalia University of Applied Sciences, Wolfenbttel, Germany
Sebastian Rudolph   Technical University Dresden, Germany

ICCS Program Commitee

Bernd Amann, France
Simon Andrews, UK
Galia Angelova, Bulgaria
Peggy Cellier, France
Dan Corbett, USA
Olivier Corby, France
Madalina Croitoru, France
Cornelius Croitoru, Romania
Licong Cui, USA
Frithjof Dau, Germany
Aldo De More, The Netherlands
Harry Delugach, USA
Juliette Dibie, France
Pavlin Dobrev, Bulgaria

Florent Domenach, Cyprus
Ged Ellis, Australia
Jérôme Euzenat, France
Catherine Faron Zucker, France
Jérôme Fortin, France
Cynthia Vera Glodeanu, Germany
Tarik Hadzic, Ireland
Ollivier Haemmerlé, France
Jan Hladik, Germany
Rinke Hoekstra, The Netherlands
John Howse, UK
Dmitry Ignatov, Russia
Mateja Jamnik, UK
Adil Kabbaj, Morocco

Mary Keeler, USA	Simon Polovina, UK
Hamamache Kheddouci, France	Uta Priss, Germany
Leonard Kwuida, Switzerland	Marie-Christine Rousset, France
Jérôme Lang, France	Sebastian Rudolph, Germany
Steffen Lohmann, Germany	Fatiha Saïs, France
Natalia Loukachevitch, Russia	Eric Salvat, France
Pierre Marquis, France	Atsushi Shimojima, Japan
Philippe Martin, France	Iain Stalker, UK
Franck Michel, France	Gem Stapleton, UK
Bernard Moulin, Canada	Gerd Stumme, Germany
Sergei Obiedkov, Russia	Nouredine Tamani, France
Peter Øhrstrøm, Denmark	Annette Ten Teije, The Netherlands
Yoshiaki Okubo, Japan	Michaël Thomazo, Germany
Nathalie Pernelle, France	Serena Villata, France
Heather D. Pfeiffer, USA	Martin Watmough, UK

Additional Reviewers

Wouter Beek	Sara Maglicane	Brigitte Safar
Jim Burton	Martin Möhrmann	Stephan Schulz
Peter Chapman	Antonio Penta	Nicolas Schwind

ICCS Administration

Proceedings

Juliette Dibie
Liliana Ibanescu
Stéphane Dervaux

Finance, Sponsors

Cassia Trojahn dos Santos

Publicity

Sven Linker

Website

Nicolas Seydoux
Rémi Cavallo

Partners

AFIA (French Artificial Intelligence Association)
INRIA (French Institute for Research in Computer Science and Automation)
Université Savoie Mont Blanc

Contents

Time Representation

Exploring the Usefulness of Formal Concept Analysis for Robust Detection of Spatio-temporal Spike Patterns in Massively Parallel Spike Trains

Alper Yegenoglu[1], Pietro Quaglio[1], Emiliano Torre[1],
Sonja Grün[1,2(✉)], and Dominik Endres[3(✉)]

[1] Institute of Neuroscience and Medicine (INM-6),
Institute for Advanced Simulation (IAS-6),
JARA BRAIN Institute I, Jülich Research Centre, Jülich, Germany
s.gruen@fz-juelich.de
[2] Theoretical Systems Neurobiology, RWTH Aachen University, Aachen, Germany
[3] Theoretical Neuroscience Group, Department of Psychology,
Philipps-University, Marburg, Germany
dominik.endres@uni-marburg.de

Abstract. The understanding of the mechanisms of information processing in the brain would yield practical impact on innovations such as brain-computer interfaces. Spatio-temporal patterns of spikes (or action potentials) produced by groups of neurons have been hypothesized to play an important role in cortical communication [1]. Due to modern advances in recording techniques at millisecond resolution, an empirical test of the spatio-temporal pattern hypothesis is now becoming possible in principle. However, existing methods for such a test are limited to a small number of parallel spike recordings. We propose a new method that is based on Formal Concept Analysis (FCA, [11]) to carry out this intensive search. We show that evaluating conceptual stability [18] is an effective way of separating background noise from interesting patterns, as assessed by precision and recall rates on ground truth data. Because of the scaling behavior of stability evaluation, our approach is only feasible on medium-sized data sets consisting of a few dozens of neurons recorded simultaneously for some seconds. We would therefore like to encourage investigations on how to improve this scaling, to facilitate research in this important area of computational neuroscience.

1 Introduction

The cerebral cortex is a highly interconnected network of nerve cells (neurons) which communicate by emitting fast electrical impulses called spikes.

A. Yegenoglu and P. Quaglio—Equal contribution; S. Grün and D. Endres—Equal contribution.

O. Haemmerlé et al. (Eds.): ICCS 2016, LNAI 9717, pp. 3–16, 2016.
DOI: 10.1007/978-3-319-40985-6_1

Millisecond-precise temporal coordination of spike emission among different neurons is believed to represent a means of information transfer and processing in the neuronal circuitry. Under this hypothesis, spatio-temporal patterns (STPs) of precisely timed spikes occur in relation to specific stimuli or behaviors.

Neural network models have been proposed that can carry out computations by the generation of STPs. The authors of [5] and later [15] proposed a model (synfire braids or polychrony, respectively) where STPs allow spiking activity to propagate throughout the network reliably. The model relies on properly set propagation delays from one neuron to the next for reliable spike production and transmission. Activity can thus propagate in the form of STPs. A special case of such networks which historically preceded them, is the synfire chain model [1,7], where propagation delays among neurons are identical and STPs formed by temporal sequences of synchronous spikes propagate from one layer of neurons to the next, due to a dense connectivity from one layer to the next. Models of this type have been shown to allow for activity propagation robustly in the presence of noise [8,27].

Evidence has been indeed provided for the occurrence of precise spatio-temporal patterns in electrophysiological data of limited size (e.g. [22,24]). However, the analysis tools used in these studies do not scale with the size of the data and thus cannot be applied to massively parallel spike trains (MPSTs) (i.e. hundreds of neurons firing in unison), as those recorded nowadays. In [29] we started to address this problem by introducing an efficient method, named SPADE (synchronous pattern detection and evaluation) to assess the presence of unexpected patterns of synchronous spikes in massively parallel spike trains. SPADE exploits to this aim frequent itemset mining techniques [6], and further assesses the statistical significance of such patterns. However, SPADE can only detect synchronous spike patterns, and not more general STPs whose individual component spikes follow each other in a temporal sequence.

We propose an approach designed to find these more general types of STPs. Compared to SPADE and other tools for the analysis of synchronous spike patterns, one difficulty lies in the increased number of patterns to look for. Adding the temporal dimension yields a number of possible patterns which is orders of magnitude larger. The occurrences of each of these patterns have to be counted, and non-chance patterns have to be differentiated from chance patterns based on properties such as the number of composing spikes or the number of pattern repetitions. Another difficulty lies in the decreased contrast between these STPs and background activity, due to the fact that the events forming the STPs are here individual rather than synchronous spikes. We deal with the first difficulty by considering only patterns that actually happened, rather than all possible patterns. More precisely, as we describe in Sect. 2.1, we study patterns that form the intent of formal concepts [11]. We address the second difficulty by evaluating the intensional stability [18] of these concepts as an indicator for the non-randomness of a concept. The application of Formal Concept Analysis (FCA) for data mining in neuroscience is not entirely new: we employed FCA for relational decoding of brain recordings [9,10]. However, FCA has not been used for the detection of STPs.

We validate the proposed method on test data consisting of parallel spike trains which comprise independent background activity, where patterns arise purely by chance, plus multiple occurrences of an STP injected in the independent data. Our method applied to these data should ideally return the injected STP, and that one only. We investigate the performance of the method in terms of true positive (TP) and false positive (FP) outcomes in a variety of scenarios, and illustrate their precision and recall properties. We also demonstrate how the computational load of the method varies for different data and analysis parameters.

We conclude by discussing benefits and current limitations of the proposed method, open problems and possible solutions.

2   Methods

2.1   FCA on Spike Data

Regarding FCA, we use the standard definitions [11]: a formal context is a triple $K = (G, M, I)$ comprising a set of formal objects G, a set of formal attributes M, and a binary relation I between the objects in G and the attributes in M. If $(g, m) \in I$ with $g \in G$, $m \in M$, we also write gIm. We denote the set of all

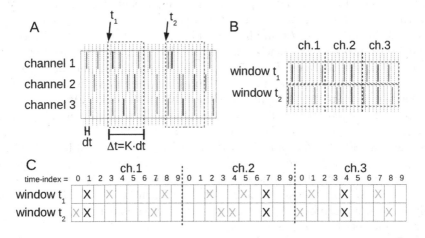

Fig. 1. *Constructing a formal context from spike recordings in parallel channels. A:* we analyze parallel neuronal recordings in multiple channels by discretizing time into bins of duration dt and chopping time windows of length $\Delta t = K \cdot dt$ out of the recorded data stream. The gray spikes are background spikes, the black spikes are members of an STP. *B:* we concatenate the recorded spike trains within a window horizontally. *C:* incidence table representation of panel B. The objects of this formal context are the time windows, indexed by their starting time. The attributes are the spike time-indexes relative to the window start, combined with the channel identities.

attributes shared by a set of objects $A \subseteq G$ as $A' = \{m \in M | \forall g \in A : gIm\}$ and likewise the set of all objects which have all the attributes in $B \subseteq M$ as $B \subseteq M$: $B' = \{g \in G | \forall m \in B : gIm\}$. A formal concept is a tuple (A, B) with extent $A \subseteq G$ and intent $B \subseteq M$ such that $A' = B$ and $B' = A$. Let $\mathfrak{B}(K)$ be the set of all concepts of K. Concepts are (partially) ordered under set inclusion on the extents: $\forall (A_1, B_1), (A_2, B_2) \in \mathfrak{B}(K) : (A_1, B_1) \leq (A_2, B_2) \Leftrightarrow A_1 \subseteq A_2$. $(\mathfrak{B}(K), \leq)$ is a complete lattice [11].

In our application, the objects are time windows within which spiking activity of neurons was observed. The example in Fig. 1A depicts two such windows labeled by their starting times t_1 and t_2. The attributes are tuples (channel-number, time-index) of the channel (or neuron) number from which a given spike was recorded, and of the spike time index relative to the window onset. Figure 1 illustrates the process of computing the relation I from spike data. Suppose we record spike trains (temporal sequences of spikes) from 3 channels, each channel recording the spikes of one neuron. We discretize the time axis into contiguous bins of duration dt. Then, we slide a time window of duration $\Delta t = K \cdot dt$ across these data, in increments of dt. dt is chosen depending on the resolution of the recording device and on the analysis needs, K is selected based on the expected maximal duration of an STP. We set $dt = 1\,\mathrm{ms}$ throughout this paper, which ensures that there is at most one spike from the same neuron in each bin. The value of K is discussed below. The contents of each window across all channels are then concatenated horizontally (see panel B). Lastly, spikes are converted to crosses, yielding the familiar incidence-table representation of I (panel C), to which we apply FCA. We use a pure Python implementation of the fast-FCA algorithm [20]. This algorithm creates the concepts in an order that embeds the usual concept order, which simplifies the subsequent evaluation of stability.

Figure 1A also shows by black ticks an STP. In this example, these spikes correspond to the attribute set $B = \{(1, 1), (2, 7), (3, 4)\}$. $\mathfrak{B}(K)$ contains a concept whose intent consists of these spikes only, plus concepts whose intent comprises time-shifted versions of these spikes at all possible times within Δt. Any of these concepts corresponds to the STP we are interested in. We arbitrarily choose the one where the first spike is aligned with the window onset. However, typically such an STP does not appear in isolation, but is embedded in background spiking activity (gray spikes). Hence, after the application of FCA, there will be many concepts which are due to these background processes, and which we wish to separate from the STP concept. We experiment with conceptual stability analysis for this purpose [18]. Specifically, we compute an intensional stability index of concepts $(A, B) \in \mathfrak{B}(K)$ [26] by

$$\sigma(A, B) = \frac{|\{C \subseteq A | C' = B\}|}{2^{|A|}} \qquad (1)$$

with the algorithm described in that paper, and filter out all concepts whose stability index is lower than a pre-fixed threshold (see Sect. 3.1). Furthermore, we filter out concepts that are time-shifted superconcepts of concepts whose stability index is higher than the threshold.

2.2   Ground Truth Data Generation

To test the performance of FCA in detecting STPs we use artificial ground truth data with controllable pattern occurrences and pattern size. We generate parallel Poisson processes to simulate independent background activity. We also generate data sets which contain STPs by injecting spike patterns into independent background activity, cf. [4,14]. The background rates are chosen to comply with the firing rates of experimental neurons, while the type of spike patterns resembles that which would be observed in data from synfire-like networks [1,5,15].

For the background activity we generate independent spike trains as realizations of 50 parallel independent Poisson processes of duration $T = 1$ s. The theoretical firing rates $r \in \{5\,\text{Hz}, 10\,\text{Hz}, 15\,\text{Hz}\}$ used to draw the processes are identical for all neurons. We generate 100 realizations for each value of r, yielding a total of 300 different background activity-only data sets. Such independent data sets are used to study the occurrence of false positives (FP), i.e. cases where patterns are detected although there were none injected.

In a second type of data set, which we use for the evaluation of true positives (TPs) and false negatives (FNs), we generate c occurrences of an STP composed of a sequence of z spikes from different neurons. To this end, we randomly select c time points $t_1, \ldots, t_c$ in the interval $[0, T - z \cdot 5\,\text{ms}]$. These times correspond to the times of the first spike of the patterns for each of the c repetitions of the pattern. Each following spike of the STP is injected 5 ms after the preceding one and into a different neuron. Therefore, the total duration of such patterns is $(z-1) \cdot 5$ ms. The pattern is chosen with regular spike delays for convenience, but this choice is not relevant for their detectability. We vary both z and c in the range $\{3, 4, \ldots, 10\}$. For each of the 64 combinations of these two parameters we generate 100 different realizations of the patterns, which we inject in an independent background activity data set generated as explained above, for each rate level r. This yields a total of $64 \cdot 100 \cdot 3 = 19200$ data sets, each containing a total of 50 neurons, z of which are involved in an STP. We then extract all concepts from each data set using FCA as explained in Sect. 2.1. Since the maximum pattern length is equal to 45 ms for $z = 10$, we fix the length K of the sliding window to 50 ms, so that the longest pattern is covered by one single window. For a data set containing a pattern with parameters (z, c), we define as a TP the correct detection of the concept (A, B), such that $A = \{t_1, \ldots, t_c\}$ and $B = \{(1, 0), (2, 5), \ldots, (z, (z-1) \cdot 5)\}$. Any other detected concept is considered as an FP. Since in each of the data sets only one pattern type is injected, the number c_{TP} of TPs per data set is either 0 or 1. Reciprocally, the number of false negatives is $c_{\text{FN}} = 1 - c_{\text{TP}}$. For a given parameter tuple (r, z, c), we determine the fraction of the 100 data sets in which we find an FP or an FN, providing us with the FP rate and the FN rate, respectively.

3   Results

3.1   Independent Data

We start evaluating the performance of the method by analyzing the occurrence of FPs in the background-only data sets. Concepts detected in these data sets are thus chance occurrences of specific spike sequences and are considered as FPs. The two parameters by which we characterize a concept (A, B) are its extent size $|A|$ and its intent size $|B|$. In terms of the spike data the extent size corresponds to the number of windows in which a specific sequence occurred, and the intent size to the number of spikes composing the sequence.

Figure 2 shows in gray code the number of concepts detected in independent data as a function of their intent size (horizontal axis) and extent size (vertical axis). We call this type of display 'pattern spectrum' in line with [12,29]. Concepts with an extent or an intent of size 1 are not counted or displayed, since single spike occurrences and non-repeating sequences are not considered as potential STPs. Each panel shows the pattern spectrum for a different neuronal firing rate r. The area of non-zero entries and thus the total number of concepts increases with the firing rate of the neurons (see also Fig. 3), which is expected since more spike sequences may occur by chance at higher firing rates due to the increased number of spikes. Most detected concepts have extent size of $|A| = 2$ or intent size of $|B| = 2$.

As mentioned above, we aim to select the true positive STPs by applying a stability analysis to the concepts. We wish to determine a stability threshold value such that the concepts that occur by chance are discarded. A suitable threshold value needs to fulfill the following constraints: (1) avoid detection of FPs, (2) have a high degree of TP detection. The histograms in Fig. 3 (left to right) correspond to the pattern spectra shown in Fig. 2 (left to right). The gray parts of the stacked bars show the counts of concepts whose intent contains only 2 spikes ($|B| = 2$) or whose extent contains only two windows ($|A| = 2$). Vice versa, the black parts correspond to the concept count for $|A| > 2$ and $|B| > 2$. Concepts with stability larger than or equal to 0.6 have always either intent or extent size equal to 2. This fact provides us with a suitable criteria for classifying all concepts which have either intent or extent size 2, or stability ≤ 0.6 as chance patterns.

3.2   Performance of Pattern Detection

Now that we have defined a suitable criteria to reject all the chance patterns composed only of spikes of the background activity (intent > 2, extent > 2, stability > 0.6), we can evaluate the performance of the method for the detection of STPs injected into artificial data, as described in Sect. 2.2. Figure 4, top row shows the number c_{FP} of FPs for data with injected patterns, and background rates $r = 5, 10, 15\,\text{Hz}$ from the left to the right panel. Each panel shows, in shades of gray, the number of data sets which contain at least one FP as a function of the number z of spikes (horizontal axis) and of the number c of occurrences of the pattern injected in the data sets (vertical axis). Because 100 simulations were carried out for each parameter set (z, r, c), the value of each entry ranges from 0 to 100. The total number of FPs is low (usually 3 or lower) but increases both with the firing rate, as we can expect from the results of the previous section, and with the pattern parameters z and c. The reason for the latter is that the larger and more frequent the injected pattern is, the more likely is the repetition

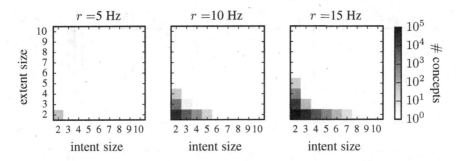

Fig. 2. *Pattern spectra of independent background-only data:* each panel shows the concept counts as a function of their intent size $|B|$ (horizontal axis) and of their extent size $|A|$ (vertical axis) detected in 100 independent data sets, composed of 50 parallel spike trains without any pattern injection mimicking background noise, for different firing rates (from left to right: $5, 10, 15\,$Hz). The number of counts is given in gray colors, as indicated by the color bar (white corresponds to 0, black to 10^5 counts).

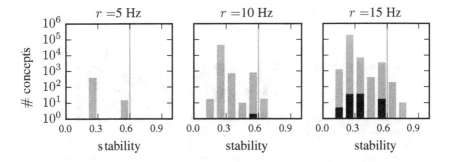

Fig. 3. *Stability histogram of the concepts in independent background-only data:* each of the three panels shows the histogram of the number of concepts found in independent background data, for different stability values. The data analyzed are the same as shown in Fig. 2, with the firing rates of the neurons changing from 5 to $15\,$Hz in the panels from left to right. The height of each bar in the histograms represents the number of concepts whose stability is in the range indicated on the horizontal axis (bin width: 0.1). The black bars (that is, the black parts of the stacked bars) show the counts of concepts with both extent *and* intent size larger than 2, whereas the gray bars show their counts when one of the two size values is equal to 2. The thin gray line indicates a threshold for the stability above which no black bars are visible, indicating that patterns with this or higher stability values have extent or intent size (or both) 2.

of spurious sequences of spikes which by chance occur with the injected pattern and which thus result in a concept with a stability index > 0.6.

In the second row of Fig. 4 we show equivalent diagrams with FN counts (c_{FN}). Each entry shows the number of data sets (out of 100 realizations) in which the injected pattern is not detected. For each rate level, the number of FNs shows a sharp decrease at a particular number of pattern injections c, independent of the pattern size z (e.g. for $r = 5\,$Hz at $c = 4$). This border increases to higher c

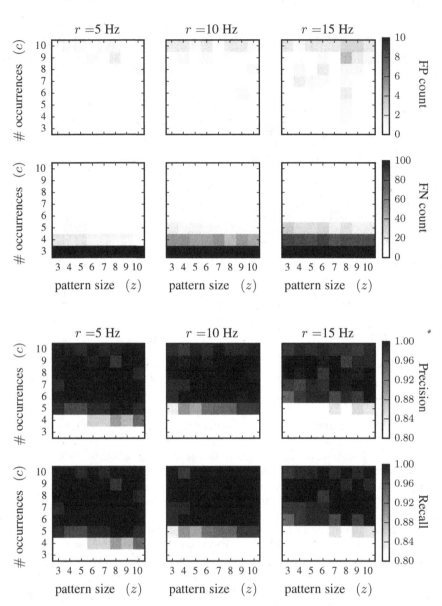

Fig. 4. *Performance of the method for various data parameters. Top half: FPs and FNs in data with injected STPs:* each panel shows the number of data sets which contain at least one FP or one FN after filtering the concepts using a stability threshold of 0.6 and a lower threshold value of 2 for both intent and extent size. The firing rates are varied from left to right from $r = 5\,\text{Hz}$ to $r = 15\,\text{Hz}$. For each display we varied along the horizontal and vertical axis the size z and the number of injections c of the injected pattern, respectively. *Bottom half: Precision and recall* corresponding to the FP/FN counts shown in the top half. The color bar for precision and recall ranges from 0.8 to 1.0, whereas all values below ≤ 0.8 are set to white.

levels for higher firing rates (from left to right). The number of FNs increases with the rates due to an increased probability of spurious superpatterns of the injected STP, causing the correct STP to be ignored by our current concept filtering procedure in favour of the superpatterns. The independence from the pattern size is a feature of intensional stability, combined with our data generation process: all windows containing the target STP are guaranteed to contain all its constituting spikes (plus possibly additional noise spikes), no matter how large z is. For all studied firing rates, patterns occurring just 3 times are never detected, but all patterns occurring 5 or more times are correctly detected, which is a consequence of our choice of the stability threshold.

To get further insight in the performance of our approach, we compute for each data set with parameters (r, z, c) its precision and recall as defined in [23]:

$$\text{precision} = \frac{c_{\text{TP}}}{c_{\text{TP}} + c_{\text{FP}}}, \text{ recall} = \frac{c_{\text{TP}}}{c_{\text{TP}} + c_{\text{FN}}}$$

We calculate the two measures separately for the same data sets used for the FP/FN evaluation. Figure 4, third row shows the average precision and recall as a function of the parameters (r, z, c). Since we are particularly interested in the high performance regime, we emphasize this range by adjusting the color code such that it covers only the interval $[0.8, 1.0]$, and we set all values ≤ 0.8 to white. Black corresponds to a value of 1.0. The performances expressed by either measure are very similar, which is due to the fact that in the high precision/high recall regime, both c_{FP} and c_{FN} are very small. Both are close to 0 for low numbers of pattern occurrences c, but drastically increase for data sets with 5 or more injections, and are basically independent of the pattern size z. This is due to the behavior of the FNs.

The acceptable tradeoff between FPs and FNs depends on the hypothesis to be tested. Since we wish to show the existence of STPs, we choose a conservative threshold guaranteeing virtually no FPs, even if that incurs some FNs.

3.3   Runtime Behavior of the FCA Algorithm

We test the runtime behavior of our method on data sets containing independent spike data without injected STPs. The presence of a few STPs would not noticeably affect the overall computational effort. Our goal is to determine whether our current implementation, which also uses the SciPy toolbox [16], is fast enough to enable the analysis of typical experimental data sets from multi-channel recordings. Since the total runtime of FCA would in general increase with the context fill ratio, we generate the data sets with a fairly high baseline firing rate of 15 Hz. We simulate 50 neurons in parallel, a number comparable to the size of modern multi-electrode recordings. We would like to be able to analyze up to 30 repeated trials of the same experiment, in chunks of 500 ms. Hence, the average number of spikes in the data is 11250. Note, however, that this number is likely to grow as experimental techniques advance.

For the profiling results shown in Fig. 5, we use a cluster with nodes consisting of 2 × Intel Xeon E5-2680v3 processors with 2.5 GHz processing speed and

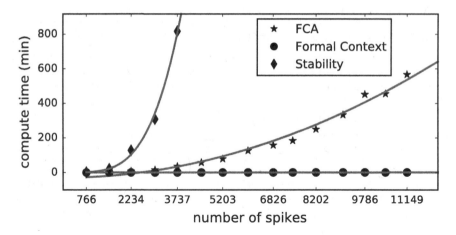

Fig. 5. *Profiling results on synthetic data: nine data sets with different number of spikes were evaluated:* The compute time to calculate the FCA (asterisk), formal contexts (filled circle), stability (diamond) and the corresponding fitted curves on each data set are shown. For details, see text.

8×16 GB DDR4 RAM, and report the time taken for context construction (circles), FCA (stars) and stability (diamonds) as a function of the total number of spikes in the data set. The number of spikes is the actual number taken from the data set used for each run. Most of the time is consumed by the computation of stability, consistent with the scaling analysis presented by [26]: exact stability evaluation time grows quadratically with the number of concepts. The curves in Fig. 5 are fits performed with a linear function for context construction, a quadratic function for FCA and thus a quartic fit for the stability. Exact stability evaluation is feasible within a day up to about 4000 spikes, corresponding to 12 experimental trials with the above-mentioned features (number of neurons, trial duration, average neuronal firing rates). Extrapolation of the runtime to 30 trials based on the quartic fit yields ≈45 days.

4 Discussion and Outlook

Information in the cerebral cortex has been hypothesized to be encoded and processed in terms of spatio-temporal patterns (STPs) of spikes generated from different neurons. Under this hypothesis, an STP is assumed to be the signature of active cell assemblies through which temporal sequences of spikes propagate at millisecond precision. Various network models, such as the synfire chain [1] and synfire braid [5,15] exist, which process information by propagation of STPs. These models are compatible with known features of biological networks, in particular in terms of neuronal connectivities. Nevertheless, they remain of a speculative nature because their existence is hard to prove. Earlier work provided evidence for the occurrence of STPs in data from small numbers of simultaneously recorded neurons [22,24]. However, these methods do not scale to massively parallel spike trains as recorded nowadays on a regular basis [25, 28]. We started to exploit a new approach based on formal concept analysis (FCA, [11]) to mine parallel spike train data for STPs involving any possible subgroup of neurons of an observed population. We use FCA for an extensive search for potential STPs, which are expressed as formal concepts. To construct the formal context a window of pre-defined duration is attached to each time step. Thus, potential STPs longer than the analysis window are not detected. Only concepts that repeat at least three times are further considered. STPs found in parallel spike trains may either occur by chance or be generated by an underlying network process, which may reoccur and thus cause repeated occurrences of the patterns. To disentangle chance from real patterns we used the stability measure introduced in [18] and explored which threshold on the stability would serve this purpose.

We tested the approach based on ground-truth data which we generated by stochastic simulations of 50 parallel spike trains. We used independent spike trains to explore the detection rate of mere chance patterns, i.e. false positives (FPs). We found that the FP rate is generally low, but increases with the neuronal firing rate (varied from 5 to 15 Hz). Most FP patterns have either low intent size (=2) or low extent size (=2). The stability distribution of these FPs revealed that chance patterns with extent *and* intent size both larger than 2 always had stability lower than 0.6. In the light of these considerations, we minimized the risk of FPs by adopting a criteria which classifies as non-chance patterns only the STPs which occur at least 3 times and are composed of at least 3 spikes, as already done for instance in [24], and additionally have stability 0.6 or larger. We then further generated data that contained STPs injected into independent background activity to investigate the amount of FPs that occurred in STP data as a result of the overlap of the injected patterns with background spikes, as well as the amount of false negatives (FNs). The rates for the background activity in these data varied again from 5 to 15 Hz, while the STPs involved 3 to 10 neurons and were injected 3 to 10 times in the data, at random time points. It turned out that FPs of this type were rare (usually less than 3 out of 100 simulations) and were always superpatterns of the injected STP. No FPs disjoint from the injected STP, i.e. composed of background spikes only, occurred. Instead, the FN rate decreased abruptly for a rate-dependent specific number of occurrences. The reason is that with more occurrences the stability of the pattern increases. The number of occurrences representing the border between high and low FN levels did not depend on the size z of the injected STP. We could draw analogous conclusions from the evaluation of the precision and recall of the method. Both quantities were low (<0.8) for a low number of pattern occurrences, but drastically increased for data sets containing patterns with 5 or more injections. Besides, both were almost independent of the pattern size z, which is due to the behavior of the FNs mentioned above. We experimented also with larger values for the stability threshold than 0.6 (up to a threshold of 0.8). Higher thresholds yielded a lower amount of FPs at the expense of higher FN levels. In the light of this trade-off, 0.6 seemed to be a suitable threshold for stability in our settings.

For potential applications of the method to experimental data, a suitable value for the stability threshold remains to be determined. Several Monte Carlo techniques exist that use independent surrogates of the original data in order to derive the statistical significance of correlations present in the original data [13]. The independent surrogates are generated from the original data by intentionally destroying the precise timing of spikes (and thus STPs as well) while preserving other features of the data (e.g. firing rates or spiking regularity) as much as possible [21]. Using such surrogate data could be an effective way to determine typical stability values of chance patterns and thus to set a suitable stability threshold. This approach however remains to be investigated. Additionally, we plan to investigate the relation between the approach suggested here to detect STPs and methods we introduced e.g. in [14,29] to detect patterns of synchronous spikes on the basis of their statistical significance. Both the stability value proposed here and statistical p-values used in other studies are meant as measures to distinguish chance and non-chance STPs. Suitable thresholds for these values must account for typical features of spiking activity, such as regularity of inter-spike intervals, temporal modulation of neuronal firing rates, rate heterogeneity across neurons, and so on.

Finally, another aspect needs to be solved before our approach becomes applicable to data of larger size than considered here (e.g. larger number of neurons, higher firing rates, longer recording time). Namely, our runtime analysis revealed that context construction and FCA are reasonably fast, but the time taken to compute the pattern stability scales as a quartic function of the number of spikes. Various steps could be undertaken to improve the computational performance of the method: a more efficient implementation of the FCA algorithm we used [20], the use of a faster algorithm [2,19], a parallel FCA algorithm [17], or an approximate rather than exact evaluation of pattern stability [3]. These improvements will be investigated in future work.

Acknowledgments. This work was partly supported by Helmholtz Portfolio Supercomputing and Modeling for the Human Brain (SMHB), Human Brain Project (HBP, EU Grant 604102), and DFG SPP Priority Program 1665 (GR 1753/4-1). DE acknowledges support from the DFG under IRTG 1901 'The Brain in Action'.

References

1. Abeles, M.: Corticonics: Neural Circuits of the Cerebral Cortex, 1st edn. Cambridge University Press, Cambridge (1991)
2. Andrews, S.: In close, a fast algorithm for computing formal concepts. In: Seventeenth International Conference on Conceptual Structures (2009)
3. Babin, M.A., Kuznetsov, S.O.: Approximating concept stability. In: Domenach, F., Ignatov, D.I., Poelmans, J. (eds.) ICFCA 2012. LNCS, vol. 7278, pp. 7–15. Springer, Heidelberg (2012)
4. Berger, D., Borgelt, C., Louis, S., Morrison, A., Grün, S.: Efficient identification of assembly neurons withinmassively parallel spike trains. Comput. Intell. Neurosci. **2010**, 1–18 (2010). doi:10.1155/2010/439648. Aricle ID 439648

5. Bienenstock, E.: A model of neocortex. Netw. Comput. Neural Syst. **6**(2), 179–224 (1995)
6. Borgelt, C.: Frequent item set mining. In: Wiley Interdisciplinary Reviews (WIREs): Data Mining and Knowledge Discovery, vol. 2, pp. 437–456. Wiley, Chichester (2012). doi:10.1002/widm.1074
7. Diesmann, M., Gewaltig, M.-O., Aertsen, A.: Characterization of synfire activity by propagating 'pulse packets'. In: Bower, J.M. (ed.) Computational Neuroscience: Trends in Research, pp. 59–64. Academic Press, San Diego (1996)
8. Diesmann, M., Gewaltig, M.-O., Aertsen, A.: Stable propagation of synchronous spiking in cortical neural networks. Nature **402**(6761), 529–533 (1999)
9. Endres, D., Adam, R., Giese, M.A., Noppeney, U.: Understanding the semantic structure of human fMRI brain recordings with formal concept analysis. In: Domenach, F., Ignatov, D.I., Poelmans, J. (eds.) ICFCA 2012. LNCS, vol. 7278, pp. 96–111. Springer, Heidelberg (2012)
10. Endres, D.M., Földiák, P., Priss, U.: An application of formal concept analysis to semantic neural decoding. Ann. Math. Artif. Intell. **57**(3–4), 233–248 (2009)
11. Ganter, B., Wille, R.: Formal Concept Analysis: Mathematical Foundations. Springer, Heidelberg (1999)
12. Gerstein, G.L., Williams, E.R., Diesmann, M., Grün, S., Trengove, C.: Detecting synfire chains in parallel spike data. J. Neurosci. Methods **206**(1), 54–64 (2012)
13. Grün, S.: Data-driven significance estimation of precise spike correlation. J. Neurophysiol. **101**(3), 1126–1140 (2009)
14. Grün, S., Abeles, M., Diesmann, M.: Impact of higher-order correlations on coincidence distributions of massively parallel data. In: Marinaro, M., Scarpetta, S., Yamaguchi, Y. (eds.) Dynamic Brain - from Neural Spikes to Behaviors. LNCS, vol. 5286, pp. 96–114. Springer, Heidelberg (2008)
15. Izhikevich, E.M.: Polychronization: computation with spikes. Neural Comput. **18**, 245–282 (2006)
16. Jones, E., Oliphant, T., Peterson, P., et al.: SciPy: open source scientific tools for Python (2001). Accessed 25 Jan 2016
17. Krajca, P., Vychodil, V.: Distributed algorithm for computing formal concepts using map-reduce framework. In: Adams, N.M., Robardet, C., Siebes, A., Boulicaut, J.-F. (eds.) IDA 2009. LNCS, vol. 5772, pp. 333–344. Springer, Heidelberg (2009)
18. Kuznetsov, S.O.: On stability of a formal concept. Ann. Math. Artif. Intell. **49**(1–4), 101–115 (2007)
19. Kuznetsov, S.O., Obiedkov, S.: Comparing performance of algorithms for generating concept lattices. J. Exp. Theoret. Artif. Intell. **14**, 189–216 (2002)
20. Lindig, C.: Fast concept analysis. In: Working with Conceptual Structures - Contributions to ICCS 2000, pp. 152–161. Shaker Verlag, August 2000
21. Louis, S., Gerstein, G.L., Grün, S., Diesmann, M.: Surrogate spike train generation through dithering in operational time. Front. Comput. Neurosci. **4**, 127 (2010)
22. Nadasdy, Z., Hirase, H., Czurko, A., Csicsvari, J., Buzsaki, G.: Replay and time compression of recurring spike sequences in the hippocampus. J. Neurosci. **19**(21), 9497–9507 (1999)
23. Olson, D.L., Delen, D.: Advanced Data Mining Techniques. Springer, Heidelberg (2008)
24. Prut, Y., Vaadia, E., Bergman, H., Haalman, I., Hamutal, S., Abeles, M.: Spatiotemporal structure of cortical activity: properties and behavioral relevance. J. Neurophysiol. **79**(6), 2857–2874 (1998)

25. Riehle, A., Wirtssohn, S., Grün, S., Brochier, T.: Mapping the spatio-temporal structure of motor cortical lfp and spiking activities during reach-to-grasp movements. Front. Neural Circ. **7**, 48 (2013). doi:10.3389/fncir.2013.00048

26. Roth, C., Obiedkov, S.A., Kourie, D.G.: On succinct representation of knowledge community taxonomies with formal concept analysis. Int. J. Found. Comput. Sci. **19**(2), 383–404 (2008)

27. Schrader, S., Grün, S., Diesmann, M., Gerstein, G.: Detecting synfire chain activity using massively parallel spike train recording. J. Neurophysiol. **100**, 2165–2176 (2008)

28. Schwarz, D.A., Lebedev, M.A., Hanson, T.L., Dimitrov, D.F., Lehew, G., Meloy, J., Rajangam, S., Subramanian, V., Ifft, P.J., Li, Z., Ramakrishnan, A., Tate, A., Zhuang, K.Z., Nicolelis, M.A.L.: Chronic, wireless recordings of large-scale brain activity in freely moving rhesus monkeys. Nat. Methods **11**, 670–676 (2014)

29. Torre, E., Picado-Muiño, D., Denker, M., Borgelt, C., Grün, S.: Statistical evaluation of synchronous spike patterns extracted by frequent item set mining. Front. Comput. Neurosci. **7**, 132 (2013)

Extracting Hierarchies of Closed Partially-Ordered Patterns Using Relational Concept Analysis

Cristina Nica[1], Agnès Braud[1]([✉]), Xavier Dolques[1], Marianne Huchard[2], and Florence Le Ber[1]

[1] ICube, University of Strasbourg, CNRS, ENGEES, Strasbourg, France
{cristina.nica,xavier.dolques,florence.leber}@engees.unistra.fr,
agnes.braud@unistra.fr
[2] LIRMM, University of Montpellier, CNRS, Montpellier, France
huchard@lirmm.fr
http://icube-sdc.unistra.fr
https://www.lirmm.fr

Abstract. This paper presents a theoretical framework for exploring temporal data, using Relational Concept Analysis (RCA), in order to extract frequent sequential patterns that can be interpreted by domain experts. Our proposal is to transpose sequences within relational contexts, on which RCA can be applied. To help result analysis, we build closed partially-ordered patterns (cpo-patterns), that are synthetic and easy to read for experts. Each cpo-pattern is associated to a concept extent which is a set of temporal objects. Moreover, RCA allows to build hierarchies of cpo-patterns with two generalisation levels, regarding the structure of cpo-patterns and the items. The benefits of our approach are discussed with respect to pattern structures.

1 Introduction

Different approaches have been designed to explore datasets containing relational data [10]. Relational Concept Analysis (RCA, [19]) classifies sets of objects described by attributes and relations, allowing the discovery of knowledge patterns and implication rules in relational datasets. RCA has been applied to various data, e.g. for software model analysis and re-engineering [2,9]. The RCA result is a family of interconnected concept lattices, where each lattice can have a huge number of concepts. Consequently, in order to facilitate the analysis step of the RCA output some special procedures for selecting relevant concepts or facilitating the navigation are compulsory.

This paper focuses on exploring qualitative temporal data using RCA, relying on its capability of classifying relational data and its hierarchical results which facilitate the analysis step. Although there exist well-known methods for mining temporal data, our aim is to explore the benefits of RCA, that structures the temporal patterns in a lattice allowing the navigation amongst them. We present

© Springer International Publishing Switzerland 2016
O. Haemmerlé et al. (Eds.): ICCS 2016, LNAI 9717, pp. 17–30, 2016.
DOI: 10.1007/978-3-319-40985-6_2

a theoretical framework for this purpose, that is based on a *temporal data model* emphasizing a *main lattice* (i.e. containing the objects of interest) from the RCA results. Any concept of the main lattice corresponds to a set of sequential patterns that is synthesized into a closed partially-ordered pattern (cpo-pattern, [6]) to ease the analysis step. Indeed, such patterns are compact, contain the same information as the sets of sequential patterns they synthesize, and are easy to interpret. On this basis, we proceed as follows (Fig. 1). Firstly, we apply RCA on a *relational context family* containing the temporal data. Secondly, we extract the sequential patterns starting from the main lattice concepts, and build the cpo-patterns. Furthermore, thanks to the hierarchical structure of the RCA output, more or less general patterns can appear, allowing the exploration of the space of cpo-patterns from common to particular trends or vice versa without extra processing being required.

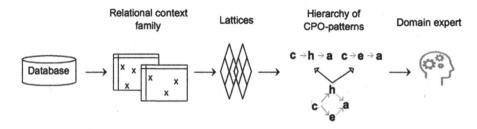

Fig. 1. Analysis process schema

The result obtained can be compared to pattern structures [14], that allow to build lattices on various data and especially graphs. As far as we know, this is still an open question to connect graph pattern structures and the set of graphs that can be built from the set of interconnected lattices of specific instantiations of RCA (like we do). To answer the question in the general case, we must consider the variety of scaling operators. Here we focus on the existential scaling operator.

The paper is structured as follows. Section 2 gives some theoretical background of our work. Section 3 introduces RCA relying on a simple temporal example. Section 4 details the RCA properties allowing to extract and to organise cpo-patterns into a hierarchy, and discusses our approach with respect to pattern structures. In addition, the whole method is described based on the same simple example. Section 5 presents related work. Section 6 concludes and gives a few perspectives of this work.

2   Background

Our approach relies both on sequential patterns and formal concept analysis domains.

2.1 Sequences, Sequential Patterns and PO-Patterns

Let $\mathcal{I} = \{I_1, I_2, \ldots, I_m\}$ be a set of *items*. An *itemset* IS is a non empty, unordered, set of items, $IS = (I_{j_1} \ldots I_{j_k})$ where $I_{j_i} \in \mathcal{I}$. Let $\mathcal{IS}$ be the set of all itemsets built from $\mathcal{I}$. A *sequence* S is a non empty ordered list of itemsets, $S = \langle IS_1 IS_2 \ldots IS_p \rangle$ where $IS_j \in \mathcal{IS}$. The sequence S is a *subsequence* of another sequence $S' = \langle IS_1' IS_2' \ldots IS_q' \rangle$, denoted as $S \preceq_s S'$, if $p \leq q$ and if there are integers $j_1 < j_2 < \ldots < j_k < \ldots < j_p$ such that $IS_1 \subseteq IS_{j_1}', IS_2 \subseteq IS_{j_2}', \ldots, IS_p \subseteq IS_{j_p}'$.

Suppose now that there is a partial order on the items, $(\mathcal{I}, \leq)$. Then the order on itemsets is defined as follows: $IS \subseteq_H IS'$ if $\forall I_j \in IS, \exists I_{j'} \in IS', I_{j'} \leq I_j$ and $j \neq k \rightarrow j' \neq k'$. The order on sequences is defined accordingly.

Sequential patterns have been defined by [1] as frequent subsequences found in a sequence database. A sequential pattern is associated to a support, i.e. the number of sequences containing the pattern. Formally, the support of a sequential pattern M extracted from a sequence database $\mathcal{D}_S$ is defined as:

$$Support(M) = |\{S \in \mathcal{D}_S | M \preceq_s S\}| \tag{1}$$

Partially ordered patterns, *po-patterns*, have been introduced by [6], to synthesise sets of sequential patterns. Formally, a *po-pattern* is a directed acyclic graph $G = (\mathcal{V}, \mathcal{E}, l)$. $\mathcal{V}$ is the set of vertices, $\mathcal{E}$ is a set of directed edges such that $\mathcal{E} \subseteq \mathcal{V} \times \mathcal{V}$, and l is a labelling function mapping each vertex to an itemset. With such a structure, we can determine a strict partial order on vertices u and v such that $u \neq v$, i.e. $u < v$ if there is a directed path from u to v. However, if there is no directed path from u to v, these elements are not comparable. Each path of the graph is a sequential pattern as defined before. The set of paths in G is denoted by $\mathcal{P}_G$. A po-pattern is associated to the set of sequences $\mathcal{S}_G$ that contain all paths of $\mathcal{P}_G$. Following Eq. 1, the support of a po-pattern is defined as:

$$Support(G) = |\mathcal{S}_G| = |\{S \in \mathcal{D}_S | \forall M \in \mathcal{P}_G, M \preceq_s S\}| \tag{2}$$

Furthermore, let G and G' be two po-patterns with $\mathcal{P}_G$ and $\mathcal{P}_{G'}$ their sets of paths. G' is a sub po-pattern of G, denoted by $G' \preceq_g G$, if $\forall M' \in \mathcal{P}_{G'}, \exists M \in \mathcal{P}_G$ such that $M' \preceq_s M$. A po-pattern G is closed, denoted cpo-pattern, if there exists no po-pattern G' such that $G \prec_g G'$ with $\mathcal{S}_G = \mathcal{S}_{G'}$.

2.2 FCA and Pattern Structures

Formal Concept Analysis (FCA, [15]) considers a formal context which is a set of objects described by attributes, and builds from it a concept lattice used to analyse the objects. Concisely, a formal context K is a 3-tuple (G, M, I), where G is a set of objects, M a set of attributes, and I the incidence relation, $I \subseteq G \times M$. $C = (X, Y)$ where $X = \{g \in G | \forall m \in Y, (g, m) \in I\}$ and $Y = \{m \in M | \forall g \in X, (g, m) \in I\}$ is a formal concept built from K. X and Y are respectively the extent and the intent of the concept. Let $\mathcal{C}_K$ be the set of all formal concepts that can be built on K. Let $C_1 = (X_1, Y_1)$ and $C_2 = (X_2, Y_2)$ be two concepts from $\mathcal{C}_K$, the concept generalisation order $\preceq_{\mathcal{C}_K}$ is here defined by $C_1 \preceq_{\mathcal{C}_K} C_2$ if and only if $X_1 \subseteq X_2$ (which is equivalent to $Y_2 \subseteq Y_1$). $\mathcal{L}_K = (\mathcal{C}_K, \preceq_{\mathcal{C}_K})$ is the concept lattice built from K. We denote by $\top(\mathcal{L}_K)$ the concept from $\mathcal{L}_K$ whose extent has all the objects, and by $\bot(\mathcal{L}_K)$ the concept from $\mathcal{L}_K$ whose intent has all the attributes.

FCA is designed to deal with binary contexts, whereas attributes can be of various forms, e.g. intervals, multi-valued, etc. To generalise the FCA approach, [14] proposed to use pattern structures. A pattern structure $(G, (D, \sqcap), \delta)$ gives a description of a set of objects G by a set of descriptions (patterns) in D, which is provided with a similarity operation $\sqcap$, such that $(D, \sqcap)$ is a meet-semilattice. $\delta : G \to D$ maps objects to their description and should verify that $\{\delta(g)|g \in G\}$ is a complete subsemilattice of $(D, \sqcap)$. Patterns can be of different types, such as vectors of intervals [16], sequences [5] or labelled graphs [14].

3   Relational Analysis of Temporal Data

We propose a general temporal modelling of sequential datasets that allows the assessment of relationships between *qualitative temporal objects*. Here, we use a toy example from the medical domain, which is illustrated in Fig. 2, to explain our general approach.

In this example, we study the symptoms (S), e.g. fever, cough and fatigue, that indicate the presence of viruses (V), e.g. influenza and hepatitis, in patients. The symptoms and viruses are detected by medical examinations (ME) and viral tests (VT), respectively. These physical examinations are identified by temporal objects *(Object, Date)* where: *Object* represents the patient, and *Date* designates the time when the physical examination was done. A patient can do several medical examinations and viral tests. Symptoms, viruses, medical examinations and viral tests are sets of objects. Viral tests are linked to viruses by some qualitative binary relations *has virus* differentiated by the type of diagnosed virus, e.g. *A*, *B* or *C*. Similarly, medical examinations are linked to symptoms by qualitative relations *has symptom* (mS or hS) differentiated by the type of identified symptoms, e.g. *moderate* or *high*. Viral tests/medical examinations and medical examinations are linked by a temporal binary relation *is preceded by* (ipb) that associates a viral test/medical examination to a medical examination if the viral test/medical examination is preceded in time by the medical examination. There is no temporal binary relation between viral tests since our aim is to study the symptoms that prognosticate distinct types of viruses. In the following, based on the relational character of the toy example, we apply RCA in order to mine these qualitative temporal data.

RCA extends the purpose of FCA to relational data. RCA applies iteratively FCA on a Relational Context Family (RCF). An RCF comprises a set $\mathcal{K}$ of object-attribute contexts and a set $\mathcal{R}$ of object-object contexts. $\mathcal{K}$ contains n object-attribute formal contexts $K_i = (G_i, M_i, I_i), i \in \{1, \ldots, n\}$. $\mathcal{R}$ contains m object-object relational contexts $R_j = (G_k, G_l, r_j), j \in \{1, \ldots, m\}$, where G_k that we call the domain of the relation and G_l that we call the range of the

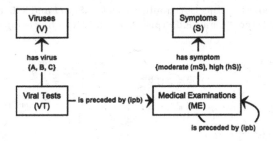

Fig. 2. The modelling of the medical sequential dataset

relation are respectively the sets of objects of K_k and K_l, and $r_j \subseteq G_k \times G_l$, with $k, l \in \{1, \ldots, n\}$. New attributes, called *relational attributes*, extend the formal contexts by using object-object relations and already created concepts. A relational attribute takes the syntactic form $qr_j(C)$, where q is a quantifier, r_j is a relation and C is a concept whose extent contains objects from the range of r_j. This paper uses the *existential* quantifier which, for a relational context $R_j = (G_k, G_l, r_j)$, creates a relation $\exists r_j$ between an object $o \in G_k$ and a concept $C = (X, Y)$ of the lattice $\mathcal{L}_{K_l}$ if $r_j(o) \cap X \neq \emptyset$. RCA process consists in applying FCA first on each object-attribute context of an RCF, and then iteratively on each object-attribute context extended by the relational attributes created using the concepts from the previous step. The RCA result is obtained when the families of lattices of two consecutive steps are isomorphic and the contexts are unchanged.

Henceforth, we try to answer the following question by means of RCA: *Can outbreaks of Influenza A virus be recognised assessing the symptoms, e.g. FEVER and COUGH, felt by patients?*

Firstly, the RCA input (RCF) is built by following the temporal data model illustrated in Fig. 2. Table 1 depicts an example of RCF on medical sequential data collected during the last year. The three tables from the left hand side represent object-attribute contexts: KS (symptoms), KVT (viral tests) and KME (medical examinations). For example, KME has no column, i.e. a medical examination is described using qualitative binary relations, and the rows represent medical examinations identified by pairs such as P1_10/01, that is a medical examination done by patient P1 on 10th of January. There is no object-attribute context of viruses due to the set of viruses that contains only *Influenza A* virus. The four tables from the right hand side represent object-object contexts: RVT-ipb-ME (viral test *ipb* medical examination), RME-ipb-ME (medical examination *ipb* medical examination), RmS (medical examination detects a moderate symptom) and RhS (medical examination detects a high symptom). For instance, RVT-ipb-ME has viral tests as rows and medical examinations as columns. A cross indicates a link between objects, e.g. the cell identified by the viral test P1_20/01 and the medical examination P1_17/01 contains a cross since both are undergone by patient P1 and 20/01 is preceded in time by 17/01.

Table 1. RCF composed of object-attribute contexts: KS, KVT and KME; temporal object-object contexts: RVT-ipb-ME and RME-ipb-ME; qualitative object-object contexts: RmS and RhS.

Object-attribute contexts | Object-object contexts

Secondly, RCA is applied on the RCF shown in Table 1 and the result is given in Fig. 3. There is a lattice for each object-attribute context: $\mathcal{L}_{KME}$ (medical examinations), $\mathcal{L}_{KS}$ (symptoms) and $\mathcal{L}_{KVT}$ (viral tests). $\mathcal{L}_{KME}$ and $\mathcal{L}_{KVT}$ are modified during the iterative steps due to the temporal and qualitative object-object contexts that have the domain KME and KVT, respectively. Each concept is represented by a box structured from top to bottom as follows: concept name, simplified intent, simplified extent. The representation of each lattice is simplified as every attribute/object is top-down/bottom-up inherited. Thus, an attribute/object is shown only in the highest/lowest concept where it appears. For example, the intent of concept CKME_4 from $\mathcal{L}_{KME}$ contains the relational attributes ∃RmS(CKS_3) and ∃RhS(CKS_3) inherited from concepts CKME_10 and CKME_9, respectively; the extent contains the objects P1_25/12 and P2_11/07 inherited respectively from concepts CKME_3 and CKME_2. The arrows represent the generalisation order. The navigation amongst these lattices follows the concepts used to build relational attributes, e.g. the aforementioned ∃RhS(CKS_3) allows us to navigate from the concept CKME_4 from $\mathcal{L}_{KME}$ to concept CKS_3 out of $\mathcal{L}_{KS}$.

4   Extracting CPO-Patterns from RCA Result

We focus here on lattices built on temporal objects. Furthermore, let us recall that we consider a main lattice, which contains the objects of interest (in the toy example, the viral tests allowing the diagnosis on a patient). In addition, concepts in these lattices contain two types of relational attributes: (1) qualitative relational attributes and (2) temporal relational attributes. The range of a qualitative attribute is a concept which represents *parameters* (e.g. virus, symptom). The range of a temporal attribute is a concept that represents *temporal observations* (e.g. medical examinations).

4.1   Proposition

In this section we give some useful properties of RCA results that help the extraction process of cpo-patterns. A temporal object in the extent of a concept from the main lattice is associated to a temporal sequence (e.g. the sequence of

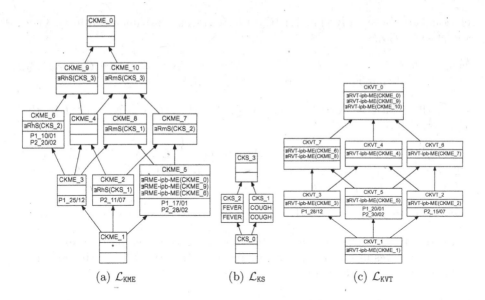

Fig. 3. The concept lattice family obtained by applying RCA on the RCF given in Table 1. The ∗ symbol represents all the relational attributes of KME

medical examinations before the final viral test). A concept extent represents thus a set of sequences. Navigating the relational attributes allows to reveal sequential patterns contained in all these sequences, as explained below.

Property 1. Each temporal relational attribute in a concept intent of the main lattice allows to extract at least one sequential pattern. On the contrary, if there is no temporal relational attribute in a concept intent, this concept represents no sequential pattern.

Indeed, let C be a concept of the main lattice and $\exists t(C_1)$ a temporal relational attribute of its intent. If C_1 intent contains a qualitative attribute $\exists q(C_2)$ then C_1 reveals an itemset of *qualitative values* (e.g. *Fever$_{high}$*); if C_1 intent contains a temporal attribute $\exists t(C_2)$ then C_1 leads to another itemset in the sequential pattern, depending on C_2 intent. The temporal relational attributes reveal the position of the itemset (in the sequential pattern). If C_1 intent contains no temporal attribute, the sequential pattern is finished.

A naive approach to extract sequential patterns out of a concept intent takes into account all its qualitative and temporal relational attributes. Nevertheless, some properties of the RCA results can be used to improve the extraction process. In the following, Properties 2 and 3 are introduced to reduce redundancy by considering only the relational attributes pointing *the most specific concepts*, and to prune temporal relational attributes that can be deduced by transitivity.

Property 2. Let $C_1 = (X_1, Y_1) \in \mathcal{L}_K$ and $C_2 = (X_2, Y_2) \in \mathcal{L}_K$ be two concepts such that $C_1 \preceq_{\mathcal{C}_K} C_2$. Let $C = (X, Y)$ be a concept which intent has two relational attributes $\exists r(C_1)$ and $\exists r(C_2)$ (derived from the same relation r). Then $\exists r(C_1) \rightarrow \exists r(C_2)$.

Proof. $\exists r(C_1) \in intent(C) \leftrightarrow \forall o \in X, r(o) \cap X_1 \neq \emptyset$. Since $C_1 \preceq_{C_K} C_2$, $X_1 \subseteq X_2$, and thus $r(o) \cap X_2 \neq \emptyset \leftrightarrow \exists r(C_2) \in intent(C)$.

Hence, the relational attributes are ordered and $\exists r(C_2)$ is redundant in the interpretation of C.

Property 3. Let t be the temporal relation. Let C, C_1 and C_2 be three concepts such that $\{\exists t(C_1), \exists t(C_2)\} \subseteq intent(C)$, and $\exists t(C_2) \in intent(C_1)$. Then $\exists t(C_2) \in intent(C)$ can be deduced from $\exists t(C_1) \in intent(C)$.

Proof. The Property 3 is directly obtained from the transitivity of the t relation.

In order to facilitate the analysis step the sets of sequential patterns extracted from the RCA result are then converted into cpo-patterns. We use therefore the pruning and merging steps proposed by [12].

Property 4. Let C be a concept of the main lattice whose intent contains at least one temporal attribute. Then C can be associated to a cpo-pattern that summarises the set of sequential patterns deriving from C. Conversely, the cpo-pattern is associated to the extent of C and its support is $|extent(C)|$.

Proof. A set of sequential patterns can be transformed into po-patterns [6]. A po-pattern associated to a concept is closed since the corresponding set of sequences is maximal or equivalently the concept extent is maximal.

Property 5. The set of cpo-patterns associated to the main lattice is ordered according to the inclusion on extents. This order corresponds to the subsumption on graphs $\preceq_g$ (cf. Sect. 2).

Proof. Let G and G' be two cpo-patterns with $\mathcal{P}_G$ and $\mathcal{P}_{G'}$ their sets of paths. Suppose G (resp. G') is associated to a concept $C = (X, Y)$ (resp. $C' = (X', Y')$) and $X \subseteq X'$. Then $Y' \subseteq Y \leftrightarrow \forall a \in Y', a \in Y$. Then $\forall M' \in \mathcal{P}_{G'}, \exists M \in \mathcal{P}_G, M' \preceq_s M \rightarrow G' \preceq_g G$.

4.2   Characterising CPO-Patterns

To characterise the items in the extracted sequential patterns, the qualitative relational attributes are analysed. To this end, we define two types of relational attributes, depending on the generality or specificity of the concept they point to.

Definition 1 (Vague/Defined Relational Attribute). *The relational attribute $\exists r(C_1)$, where C_1 is a concept of lattice $\mathcal{L}_K$, is called Vague if $C_1 \equiv \top(\mathcal{L}_K)$, respectively it is Defined if $C_1 \prec_{C_K} \top(\mathcal{L}_K)$.*

Relying on the taxonomy of items revealed by RCA, we define three types of items that disclose abstract and concrete information from the analysed data as follows:

- let C be a concept whose intent has no qualitative relational attribute, e.g. CKME_0, see Fig. 3. Then the extracted item is an **abstract** item, denoted by "$?_?$". The abstract item describes a collection of objects which point out the occurrence of dissimilar parameters having dissimilar qualities;
- let C be a concept whose intent contains a vague qualitative relational attribute $\exists quality(\top)$ (e.g. CKME_10). The extracted item is an **abstract qualitative** item, denoted by "$?_{quality}$". The abstract qualitative item describes a collection of objects which point out the occurrence of dissimilar parameters having the same quality;
- let C be a concept whose intent contains a defined qualitative relational attribute $\exists quality(C_1)$, with $extent(C_1) = \{value\}$ (e.g. CKME_7). The extracted item is a **concrete qualitative** item, denoted by "$value_{quality}$". The concrete qualitative item describes a collection of objects which point out the occurrence of the same concrete parameter having the same quality.

These types of items allow us to define the partial order $(\mathcal{I}, \leq)$ on the extracted items. For every *value* from the parameter set and every *quality* from the qualitative relations the relation $\leq$ is defined as follows:

- $value_{quality} \leq ?_{quality}$
- $?_{quality} \leq ?_?$
- $value_{quality} \leq ?_?$
- $value_{quality} \leq value_{quality}$.

The various extracted items allow us to introduce three new types of cpo-patterns.

Definition 2 (Abstract/Hybrid/Concrete cpo-pattern). *A cpo-pattern is as follows:*

- *Abstract if it contains only abstract and/or abstract qualitative items;*
- *Hybrid if it contains abstract and/or abstract qualitative items and concrete qualitative items;*
- *Concrete if it contains only concrete qualitative items.*

Hybrid patterns can be characterised using a measure of precision referred to as *accuracy*.

Definition 3 (Accuracy(v)). *Let G be a cpo-pattern and $\mathcal{I}_G$ the multiset of items labelling the nodes of G ($\forall I \in \mathcal{I}_G$, $I \in \mathcal{I}$). Let $\mathcal{I}_G^c$ be the subset of $\mathcal{I}_G$ containing the concrete qualitative items. The accuracy of G is defined as the ratio of the number of items in $\mathcal{I}_G^c$ to the total number of items in $\mathcal{I}_G$.*

$$v(G) = \frac{|\mathcal{I}_G^c|}{|\mathcal{I}_G|} \in [0, 1] \qquad (3)$$

If G is abstract, $v(G) = 0$; if G is concrete $v(G) = 1$.

4.3   RCA Based CPO-Patterns vs. Pattern Structures

In a previous work [12], cpo-patterns were directly extracted from sequences, using a sequential pattern mining algorithm. In these cpo-patterns, there is no order on the set of items $\mathcal{I}$, i.e. there are only concrete items. The resulting set of cpo-patterns, D, with the intersection operation on graphs $\sqcap$ allows to build a pattern structure $(G, (D, \sqcap), \delta)$ where G is the set of objects described by the cpo-patterns, through the δ relation.

The resulting lattice can be compared to the hierarchy of cpo-patterns built in our RCA-based approach. Firstly, in our approach, the hierarchy is built directly from the RCA result. Secondly our approach produces a hierarchy of items and thus more general cpo-patterns than in [12]. Finally, the RCA-approach allows both to navigate along the sequences and to synthesize them within cpo-patterns.

4.4   Implementation with the Toy Example

To illustrate our method, let us examine the concept CKVT_6 of $\mathcal{L}_{KVT}$ (Fig. 3(c)). Its intent contains four temporal relational attributes, the most specific being ∃RVT-ipb-ME(CKME_7) and ∃RVT-ipb-ME(CKME_9). On the contrary, the intents of concepts CKME_7 and CKME_9 (Fig. 3(a)) contain no temporal relational attribute. Accordingly, the chains of concepts starting from CKVT_6 are ⟨(CKVT_6)(CKME_7)⟩ and ⟨(CKVT_6)(CKME_9)⟩ as shown in Fig. 4. From these chains two sequential patterns can be extracted, denoted respectively $S1_{\text{CKVT\_6}}$ and $S2_{\text{CKVT\_6}}$.

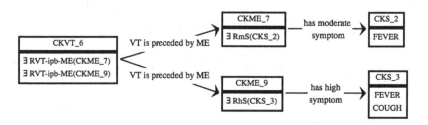

Fig. 4. The set of concept intents/extents used to interpret the concept CKVT_6

Relying on the aforementioned types of items, the itemsets of $S1_{\text{CKVT\_6}}$ and $S2_{\text{CKVT\_6}}$ can be defined. Both extracted chains of concepts begin with CKVT_6 that by default represents the concrete qualitative item "$Influenza_A$". The intent of concept CKME_7 contains both types of qualitative relational attributes: the vague ∃RmS(CKS_3) and the defined ∃RmS(CKS_2). CKS_2 $\prec_{C_{KS}}$ CKS_3 (Fig. 3(b)), therefore the interpretation of CKME_7 is based on ∃RmS(CKS_2) that represents the concrete qualitative item "$FEVER_{moderate}$". Accordingly, the extracted sequential pattern is $S1_{\text{CKVT\_6}} = \langle(\text{Influenza}_A)(\text{FEVER}_{moderate})\rangle$. Following the same principle, the intent of CKME_9 has only the vague qualitative relational attribute ∃RhS(CKS_3) that represents the abstract qualitative item "$?_{high}$". Consequently,

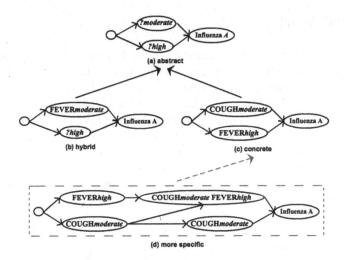

Fig. 5. An excerpt from the hierarchy of cpo-patterns obtained for the RCF out of Table 1. The (d) cpo-pattern is extracted by adding new medical examinations to the RCF

the extracted sequential pattern is $S2_{\text{CKVT\_6}} = \langle(\text{Influenza}_A)(?_{\text{high}})\rangle$. The full set of concept intents and extents navigated by our approach is shown in Fig. 4. The cpo-pattern summarising the two sequential patterns is depicted in Fig. 5(b).

Following our method, i.e. extracting one cpo-pattern for each concept of the main lattice (Fig. 3(c)), a hierarchy of cpo-patterns is obtained. In Fig. 5, the abstract, hybrid and concrete cpo-patterns represent respectively the intent of concepts CKVT_0, CKVT_6 and CKVT_7. The abstract cpo-pattern (Fig. 5(a)) subsumes a group of cpo-patterns that share the less accurate common trend: *often before Influenza A virus patients felt in any order a high symptom and another moderate symptom*. The hybrid cpo-pattern (Fig. 5(b)) subsumes a subgroup of the aforementioned group, i.e. it is a specialisation of the abstract pattern (Fig. 5(a)). This subgroup encapsulates individual cpo-patterns that share the more or less accurate common trend: *less often before Influenza A virus patients felt in any order moderate FEVER and another high symptom*. The concrete cpo-pattern (Fig. 5(c)) is another specialisation of the abstract pattern (Fig. 5(a)). This pattern subsumes a subgroup of individual cpo-patterns that share the accurate common trend: *less often before Influenza A virus patients felt in any order moderate COUGH and high FEVER*. The concrete cpo-pattern depicted in Fig. 5(d) is extracted if new medical examinations are added to the RCF from Table 1. This pattern highlights the two generalization levels of the extracted hierarchy. Firstly, the structure of this pattern is more specific than the structure of the ancestors of the pattern, i.e. this pattern contains more vertices and more edges. Secondly, the generalisation of items is illustrated. For instance, the pattern reveals the rule $\{COUGH_{moderate}, FEVER_{high}\} \Rightarrow \{Influenza_A\}$ that is a specialisation of the rule $\{FEVER_{high}\} \Rightarrow \{Influenza_A\}$ revealed by Fig. 5(c).

5   Related Work

To our knowledge, this is the first time that RCA is used to explore sequential datasets. There are, however, various related FCA approaches, e.g. [22] introduced Temporal Concept Analysis where objects are characterised with a date and a state (i.e. a set of attributes). Data are merged into a single context, and the resulting concept lattice is analysed thanks to the date element in the concepts, so that temporal relations between concepts are actually revealed by the analyst. This approach has been used to analyse sequential data about crime suspects [18]. In our RCA approach, the temporal relation between dates is considered as an object-object relation and it links concepts from several lattices. In [13], sequential datasets are processed without involving any partial order. In [6], closed subsequences are mined and then grouped in a lattice similar to an FCA lattice. In [4], sequential data are mapped onto pattern structures whose projections are used to build a pattern concept lattice. The authors combine the stability of concepts and the projections of pattern structures in order to select relevant patterns.

Besides, there exist various methods to explore qualitative sequential data. Indeed, sequential pattern mining is an active research area, in relation to the exponential growth of temporal and spatio-temporal databases. Sequential patterns have been introduced by [1] and used for different purposes, e.g. classification [7] or prediction [21]. Such an approach has been developed within the Fresqueau project and focused on cpo-patterns, which were selected through various measures [11]. Sequential pattern mining approaches are more efficient from a scalability point of view, but RCA enables to deepen the result analysis by navigating within the lattice family. Moreover it reveals a taxonomy from the data, that can be used to organise the cpo-patterns. Such results can be related to [20], where generalized sequential patterns are extracted based on a user-given taxonomy. RCA allows to discover this taxonomy.

Some authors proposed to combine RCA and pattern structures. In [8] RCA is adapted to integrate a description of G_1, a set of source objects with descriptors (coming from a pattern structure $(G_1, (D, \sqcap), \delta)$) and relational attributes to a set of concepts on a target formal context (G_2, M_2, I_2). The relational attributes are built for a relation $r \subseteq G_1 \times G_2$ and the usual scaling operators, like $\exists$ or $\forall\exists$. This is formalized as a "heterogeneous pattern structure". An application to Information Research domain is described, where source objects are documents, descriptors are vectors of intervals of LV values, target objects are terms grouped into concepts when they have same meaning (represented by a synset), and the relation r connects documents to their included terms. LV are Latent Variables that abstract hidden topics spread over the documents. Finally, there is no extraction of complex graph patterns as in our case, since there is only one relation and descriptors in the initial pattern structure are interval vectors.

6 Conclusion

In this paper, we have presented a theoretical framework for exploring temporal data using RCA. Our work proposes a comprehensive process for exploring sequential datasets which spans: (1) the relational analysis step that relies on a temporal data model which allows to emphasize the objects of interest in the study and (2) the extraction step of cpo-patterns from the RCA result.

The result is a hierarchy of cpo-patterns associated to sets of temporal objects, that can be compared to a lattice of pattern structures. With respect to pattern structures, the proposed approach, thanks to RCA, allows both to navigate along the sequences and to build a set of cpo-patterns including various levels of generalisation.

Our method was applied to sequential datasets, dealing with biological and physico-chemical parameters sampled in waterbodies [3]. Data were collected from french databases during the ANR 11 MONU 14 Fresqueau project. Results showing the effectiveness of our approach on large datasets are presented in [17] and are available at a website[1].

In the future, we will test our approach on other relational data such as spatial data, where various relations can be considered together (topology, distance, orientation).

References

1. Agrawal, R., Srikant, R.: Mining sequential patterns. In: International Conference on Data Engineering, pp. 3–14 (1995)
2. Arévalo, G., Falleri, J.-R., Huchard, M., Nebut, C.: Building abstractions in class models: formal concept analysis in a model-driven approach. In: Wang, J., Whittle, J., Harel, D., Reggio, G. (eds.) MoDELS 2006. LNCS, vol. 4199, pp. 513–527. Springer, Heidelberg (2006)
3. Berrahou, L., Lalande, N., Serrano, E., Molla, G., Berti-Équille, L., Bimonte, S., Bringay, S., Cernesson, F., Grac, C., Ienco, D., Le Ber, F., Teisseire, M.: A quality-aware spatial data warehouse for querying hydroecological data. Comput. Geosci. Part A **85**, 126–135 (2015)
4. Buzmakov, A., Egho, E., Jay, N., Kuznetsov, S.O., Napoli, A., Raïssi, C.: On mining complex sequential data by means of FCA and pattern structures. Int. J. Gen. Syst. **45**, 135–159 (2016)
5. Buzmakov, A., Egho, E., Jay, N., Kuznetsov, S.O., Napoli, A., Raïssi, C.: FCA and pattern structures for mining care trajectories. In: Proceedings of the International Workshop FCA4AI at IJCAI 2013. CEUR Workshop Proceedings, vol. 1058, pp. 7–14. CEUR-WS.org (2013)
6. Casas-Garriga, G.: Summarizing sequential data with closed partial orders. In: 2005 SIAM International Conference on Data Mining, pp. 380–391 (2005)
7. Cheng, H., Yan, X., Han, J., Hsu, C.: Discriminative frequent pattern analysis for effective classification. In: International Conference on Data Engineering, pp. 716–725 (2007)

[1] http://icube-sdc.unistra.fr/en/img_auth.php/c/c4/Mining_Hydroecological_Data_using_RCA.pdf.

8. Codocedo-Henriquez, V.: Contributions to indexing and retrieval using formal concept analysis. Doctoral thesis, Université de Lorraine, September 2015
9. Dolques, X., Huchard, M., Nebut, C., Reitz, P.: Fixing generalization defects in UML use case diagrams. Fundam. Inform. **115**(4), 327–356 (2012)
10. Džeroski, S.: Relational data mining. In: Maimon, O., Rokach, L. (eds.) Data Mining and Knowledge Discovery Handbook, pp. 869–898. Springer, New York (2005)
11. Fabrègue, M., Braud, A., Bringay, S., Grac, C., Le Ber, F., Levet, D., Teisseire, M.: Discriminant temporal patterns for linking physico-chemistry and biology in hydro-ecosystem assessment. Ecol. Inform. **24**, 210–221 (2014)
12. Fabrègue, M., Braud, A., Bringay, S., Le Ber, F., Teisseire, M.: Mining closed partially ordered patterns, a new optimized algorithm. Knowl.-Based Syst. **79**, 68–79 (2015)
13. Ferré, S.: The efficient computation of complete and concise substring scales with suffix trees. In: Kuznetsov, S.O., Schmidt, S. (eds.) ICFCA 2007. LNCS (LNAI), vol. 4390, pp. 98–113. Springer, Heidelberg (2007)
14. Ganter, B., Kuznetsov, S.O.: Pattern structures and their projections. In: Delugach, H.S., Stumme, G. (eds.) ICCS 2001. LNCS (LNAI), vol. 2120, pp. 129–142. Springer, Heidelberg (2001)
15. Ganter, B., Wille, R.: Formal Concept Analysis: Mathematical Foundations. Springer, Heidelberg (1999)
16. Kaytoue, M., Assaghir, Z., Messai, N., Napoli, A.: Two complementary classification methods for designing a concept lattice from interval data. In: Link, S., Prade, H. (eds.) FoIKS 2010. LNCS, vol. 5956, pp. 345–362. Springer, Heidelberg (2010)
17. Nica, C., Braud, A., Dolques, X., Huchard, M., Le Ber, F.: L'analyse relationnelle de concepts pour la fouille de données temporelles - Application à l'étude de données hydroécologiques. Revue des Nouvelles Technologies de l'Information Extraction et Gestion des Connaissances, EGC 2016, RNTI-E-30, pp. 267–278 (2016)
18. Poelmans, J., Elzinga, P., Viaene, S., Dedene, G.: A method based on temporal concept analysis for detecting and profiling human trafficking suspects. In: Artificial Intelligence and Applications, AIA 2010, pp. 1–9 (2010)
19. Rouane-Hacene, M., Huchard, M., Napoli, A., Valtchev, P.: Relational concept analysis: mining concept lattices from multi-relational data. Ann. Math. Artif. Intell. **67**(1), 81–108 (2013)
20. Srikant, R., Agrawal, R.: Mining sequential patterns: generalizations and performance improvements. In: Apers, Peter M.G., Bouzeghoub, Mokrane, Gardarin, Georges (eds.) EDBT 1996. LNCS, vol. 1057, pp. 3–17. Springer, Heidelberg (1996)
21. Wang, M., Shang, X., Li, Z.: Sequential pattern mining for protein function prediction. In: Tang, C., Ling, C.X., Zhou, X., Cercone, N.J., Li, X. (eds.) ADMA 2008. LNCS (LNAI), vol. 5139, pp. 652–658. Springer, Heidelberg (2008)
22. Wolff, K.E.: Temporal concept analysis. In: ICCS 2001 Workshop on Concept Lattice for KDD, 9th International Conference on Conceptual Structures, pp. 91–107 (2001)

The Interpretation of Branching Time Diagrams

David Jakobsen[(⊠)] and Peter Øhrstrøm

Department of Communication and Psychology, Aalborg University,
Rendsburggade 14, 9000 Aalborg, Denmark
{davker, poe}@hum.aau.dk

Abstract. The use of branching time diagrams in tense logic was originally suggested in 1957 by Saul Kripke. During the following years, A.N. Prior (1914–1969) developed models with forwards branching and backwards linearity. Prior's work on tense logic inspired several logicians during the 1960s, such as Nino Cochiarella (born 1933) and Henrik von Wright (1916–2003). Both of them questioned Prior's idea of a linear past. In the present paper, we argue that the best way to take advantage of the results of the various discussions of branching time since Prior would be to make use of a Molinistic version of the Ockhamistic model. We argue that this interpretation of branching time can reflect and support natural language reasoning in a very useful manner.

Keywords: Branching time · A.N. Prior · Time and existence · Tense logic · Concept of time · Natural language reasoning

1 A.N. Prior on Tense Logic

A.N. Prior (1914–1969) founded modern tense logic in the early 1950s, and it is an important approach to systems handling natural language reasoning. The basic idea in tense logic is the use of the operators F ('it will be that'), P ('it has been that'), H ('it has always been that') and G ('it is always going to be that'). Some systems also include a possibility operator $\Diamond$ ('it is possible that'). Prior was highly interested in the representation of natural language reasoning in terms of this formal language.

Some of Prior's colleagues criticized Prior's logic, like the Australian philosopher J.J.C. Smart. While Prior was convinced that time is dynamic, Smart held the opposite, which he had argued the year before they met in 'The River of Time' [13]. In a letter to Prior Jack Smart warned Prior against the use of tense logic:

> I still get the feeling that these lectures would more clearly represent your genius if you cut down on the metaphysics and stepped up the logic. As far as I can see you are at present trying to formulate ordinary tense logic. ... This produces a pretty cumbersome system. That is, not a pretty system! (The tenseless logic is far superior aesthetically) ... I would, honestly, strongly suggest cutting down on the quasi-metaphysics and increase the amount of formal logic [14].

Prior disagreed. According to him, the logician should give the 'metaphysician, perhaps even the physicist, the tense logic that he wants, provided that it be consistent' [11:59]. Prior demonstrated that his tense-logic could compete with the best logical languages in terms of consistency and rigour.

© Springer International Publishing Switzerland 2016
O. Haemmerlé et al. (Eds.): ICCS 2016, LNAI 9717, pp. 31–39, 2016.
DOI: 10.1007/978-3-319-40985-6_3

2 Saul Kripke and the Notion of Branching Time

One of the first to respond to Prior's book *Time and Modality* [9] was the young 17-year-old Saul Kripke. In a letter dated 3 September, 1958, Kripke suggested a new model for representing the concept of time related to Prior's discussion of indeterminism [12, 16]. This model was, in fact, the very first presentation of the idea of branching time in logic (see Fig. 1).

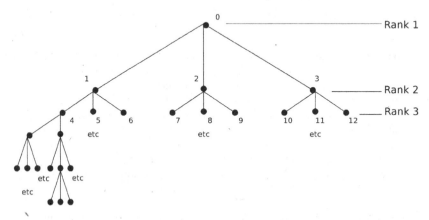

Fig. 1. Kripke's diagrammatical presentation of branching time in a letter to Prior in 1958.

In Kripke's model, 1, 2 and 3 are possible futures, directly accessible from the present moment, 0. The model also contains "future futures" as well as counterfactual moments. From every future point in the model, a new subtree exists, 'consisting of its own present and future' [6]. Prior accepted the idea of branching time since he saw it as a nice representation of natural language reasoning. Formally, the system of branching time should be conceived as a temporal structure (TIME, <), where TIME is the set of temporal moments, and < is a partial ordering of the moments. We define chronicles as linear and maximal subsets of TIME. In terms of Fig. 1 the line from 0 to 3 and 11 (and further on) would be an example of a chronicle.

Truth, in this context, is conceived as a function, π, defined on TIME $\times$ Φ, where Φ is the set of propositional variables on which propositional variables of the logical system can range. This means that for any pair (t,q) of a temporal moment and a propositional constant of the logical language there is a truth value $\pi(t,q)$ as either 0 (false) or 1 (true). In Kripke's original system, truth is only related to the elements of TIME (i.e., the moments). In this case, the truth condition for the proposition $F\varphi$ can simply be written in this way:

$t \models F\varphi$ if there is a t' with $t < t'$, such that $t' \models \varphi$

As we shall see in the following section, the valuation of propositions may in some systems be related not just to moments, but also to chronicles. This means that we may have to speak about truth at a temporal moment t, on a chronicle c, to which t belongs.

3   Linear Time Versus Branching Time

Prior's publication of *Time and Modality* [9] launched a decade of research into tense logic. One of the important writers from this period was Nino Cocchiarella (born 1933), who wrote his Ph.D. thesis in 1965 on tense logic. Cocchiarella found great inspiration in Prior's work. Like Prior, he wanted a tense logic corresponding to common sense reasoning.

Cocchiarella's tense logic, just as Prior's, may be conceived as an investigation of a part of reality. The move away from the earlier period of anti-metaphysics also is visible in Cocchiarella's treatment of logical notions, like transitivity. In fact, he claimed that "it is clear that we are not dealing with the structure of time if the earlier-than relation is not assumed to be transitive" [5:39].

Cocchiarella's logic included the axioms:

(1) $(Pp \land Pq) \supset P(p \land q) \lor P(p \land Pq) \lor P(q \land Pp)$
(2) $(Fp \land Fq) \supset F(p \land q) \lor F(p \land Fq) \lor F(q \land Fp)$

These axioms serve to secure backwards and forwards linearity, respectively. At least it is evident that the following branching time diagram illustrates a denial of axiom (2), since it shows that Fp and Fq can both be true at the moment t_2 (or earlier) without $F(p \land q)$, $F(p \land Fq)$, or $F(q \land Fp)$ being true at that moment (Fig. 2).

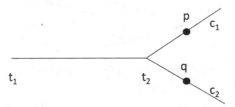

Fig. 2.   Cocchiarella's axiom 2 is not compatible with a branching view on time.

Prior was interested in Cocchiarella ideas and worked with his tense-logical formulations of backward and forward linearity. Prior was able to prove that these properties of linearity can be expressed alternatively in terms of the axioms [15:207]:

(1') $FPq \supset (Pq \lor q \lor Fq)$
(2') $PFq \supset (Pq \lor q \lor Fq)$

Clearly, axioms (1') and (2') appear to be much more elegant than (1) and (2). However, Prior preferred models that are forwards branching, but backwards linear.

In Fig. 3, *a* and *d* are world-state propositions (sometimes called instant-proposition). According to Prior, "A world-state proposition in the tense-logical sense is simply *an index of an instant*... In this sense of 'instant' it is a tautology that a world-proposition is true at one instant only (it is true only when *that* world-proposition is true) and so it is as strong as any proposition that is ever can be" [11:188–189]. Clearly, an instant-proposition is only true once. For this reason, instant-propositions

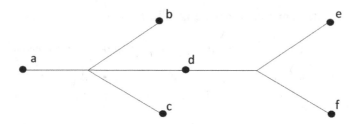

Fig. 3. Prior presented this diagram in [11:190].

are very useful for the tense-logical description of the properties of branching time diagrams.

Cocchiarella considered models of time with divergent paths into the past as well as into the future. Prior was interested in what this could mean. Actually the question was brought up in his correspondence with von Wright. In a letter to von Wright dated 1 February 1966 Prior considered the following diagram that allows alternative pasts (Fig. 4):

Fig. 4. A branching time diagram from Prior's correspondence with von Wright.

In his discussion of the idea of branching time in [11], Prior considered the diagram (Fig. 5):

Fig. 5. A branching time diagram with an ultimate future [11:28]

Prior rejected models of this kind. If there is going to be just one ultimate future, it cannot have different pasts. He claimed that "we cannot say that it will all be the same in a hundred years' time, nomatterwhathappensinbetween;sinceonethingthatwillbedifferentwillbewhat, by then, has been the case" [11:29].

It is important to stress that there is a significant difference between an epistemological and an ontological approach to the problem of alternative pasts. From an epistemological point of view, a historian, of course, may refer to the uncertainty concerning the past. We do not know exactly what happened. There are many open problems regarding the past. But although we do not know whether the proposition

'Joe had a beer yesterday at noon' is true or false, we are sure that the proposition is either true or false seen from an ontological point of view. In terms of instant-propositions, the point is that any such proposition implies anything that is true at the instant in question. Given that the instant-proposition includes a complete history of everything that was the case, the instant can only have one past.

4   The Peircean Interpretation of Branching Time

According to Prior's so called Peircean interpretation of time [11:128ff], the attribution of a truthvalue in the semantic model is laid down by the recursive definitions:

$t \models q$ if q is a propositional constant with $\pi(t,q) = 1$
$t \models \sim\varphi$ if it is not the case that $t \models \varphi$
$t \models F\varphi$ if for any c with $t \in c$ there is a $t' \in c$ with $t < t'$, such that $t' \models \varphi$
$t \models P\varphi$ if for any c with $t \in c$ there is a $t' \in c$ with $t' < t$, such that $t' \models \varphi$

In some cases, we also would like to have the temporal metric incorporated as the operators $F(n)$ ('in n days it will be the case that ...') and $P(n)$ ('n days ago it was the case that ...').[1] The truth conditions for such operators are straightforward. Given this model some common assumptions in natural language reasoning have to be rejected. One such example is illustrated in Fig. 6 below:

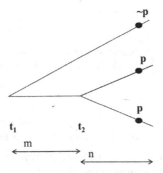

Fig. 6. The Peircean model of branching time violates $q \supset P(m)F(m)q$ with $F(n)p$ for q, since $F(n)p$ is true at t_2, whereas $P(m)F(m)q$ $F(n)p$ cannot be verified at that moment.

A further consequence of this model is that $(Fp \lor F\sim p)$ is not true in general, which certainly may be said to be a problem from a common sense and natural language point of view. A Peircean model must also make a distinction between $F(n) \sim p$ and $\sim F(n)p$, which is clearly also odd when seen from a natural language point of view. Finally, the Peircean will have to identify 'tomorrow' with 'necessarily tomorrow'. All this makes the Peircean model unattractive from a natural language perspective.

---

[1] The use of $F(n)p$ and $P(n)p$ does not presuppose anything regarding discreteness of time.

5 The Ockhamistic Interpretation of Branching Time

While the Peircean model of time was preferred by Prior, he also worked with an Ockhamistic representation of time [11:122]. Truth values, in the Ockhamistic model, can be laid down by the recursive definitions:

$t, c \models q$ if q is a propositional constant with $\pi(t,q) = 1$

$t, c \models \sim\varphi$ if it is not the case that $t, c \models \varphi$

$t, c \models F\varphi$ if there is a $t' \in c$ with $t < t'$, such that $t', c \models \varphi$

$t, c \models P\varphi$ if there is a $t' \in c$ with $t' < t$, such that $t', c \models \varphi$

Again, a metric can be added in a straightforward manner. The crucial property of the Ockhamistic model is that, here, truth at a moment, t, depends on the choice of a chronicle through t. This property can be illustrated by the diagram in Fig. 7.

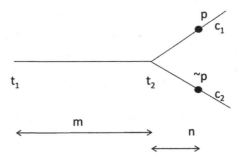

Fig. 7. The Ockhamistic model of branching time where truth is relative to a moment on a chronicle. $F(n)p$ is true at t_1 relative to the chronicle c_1, whereas $F(n)\sim p$ is true at t_2 relative to the chronicle c_2.

In the Ockhamistic model, we introduce a primitive possibility operator, $\Diamond$:

$t, c \models \Diamond\varphi$ if there is a chronicle c' with $t \in c$ 'such that $t, c' \models \varphi$

Using this modal operator, it becomes clear that from an Ockhamistic point of view, there is a distinction between the three expressions $\Diamond F(n)q$, $\Box F(n)q$, and $F(n)q$, where $\Box$ is defined as $\sim\Diamond\sim$. This is obviously attractive from a common sense and a natural language point of view. Except from two objections, the Ockhamist model, in general, reflects natural language reasoning in a nice manner. One objection to the model is that the truth value of certain past tense propositions about the future like $P(m)F(m)F(n)p$ will depend on the chronicle. In a letter to Kripke, Prior states, "a thing can only be undetermined if it is not-yet-past" [10]. According to Prior, the moment a thing is past, it is now unpreventable. If the Ockhamist view is true, though, the present will keep not only the future open but also future dependent truth about the past. Another objection is that the Ockhamist model has no simple notion of truth at a moment. That is a far shot from the natural language, which is an important reason for accepting tensed logic in the first place. Asked by someone if we will visit him, it would be weird to answer, "Yes, relative to a certain chronicle we will". In common sense reasoning we normally assume that the contingent future can be true now. This is in fact also what Ockham himself held [17].

6   The Molinistic Interpretation of Branching Time

The ambition of a Molinistic interpretation of branching time is to restore the unrestricted notion of truth at a moment. The system is inspired by the writings of the Jesuit monk Luis de Molina (1535–1600). Although he neither made use of symbolic logic nor any kind of graphical models, Molina developed a system involving the idea of a true future based on the doctrine of divine foreknowledge. According to Molina, God knows the truthvalue of any future contingent proposition even in the counterfactual case (known as 'middle knowledge'). Molina claimed that, in some cases, natural language reasoning assumes middle knowledge. His argument was based on a certain passage from the Bible, 1 Kings 23:10–12. In this text David is told by God, that if he enters Keila, Saul will choose to come to the city, and the citizens of Keila will choose to hand David over to Saul. Molina concludes that God knows "two future contingents, which depended on human choice, and He revealed them to David. Yet they never have existed and never will exist in reality" [8].

The semantics of this natural language reasoning requires a notion like branching time with a selected future at actual as well as counterfactual moments. The following branching time diagram indicates this semantics with arrows (Fig. 8):

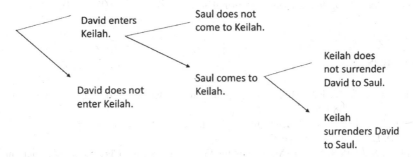

Fig. 8. A Molinistic branching time diagram corresponding to 1 Kings 23:10–12 in the Old Testament. The text from narrates a true future where David does not enter Keilah as well as a true counterfactual future, which would have been true, had David entered Keilah.

From a natural language perspective, it is attractive that the Molinistic model provides a semantics for counterfactual statements like 'it would have been' and 'it could have been the case'. Representation of common sense notions is furthermore important for natural language processing, as argued by Liu and Singh "because the implicit conceptual framework of human language makes nodes and symbols meaningful by default, gives us a way to quantify the similarity of nodes and symbols, and is thus more amenable to inductive reasoning" [7].

The attractive qualities of the Molinistic model are also found in secular examples (see [2:379]). In the modern discussion of this model Belnap and Green have used the term 'the thin red line'. However according to a Molinistic interpretation the thin red line is actually a function TRL, from TIME to the set of chronicles. TRL(t) is the selected chronicle through the moment, t. Using this function, we may define truth in a Molinistic model in terms of the approach applied in the Ockhamistic model. We may simply say that a proposition is true at an instant t in the Molinistic sense if and only if it is true at t in the Ockhamistic sense for the chronicle TRL(t) [2, 3, 17].

Nuel Belnap et al. have accepted that the Molinistic solution is possible. However, they have argued that the TRL represents 'a mysterious realm of fact' [3:169]. A defence of the metaphysics of the TRL recently has been undertaken by Borghini and Torrengo [4].

7   Conclusion

As we have seen, there are several issues to consider if we want to establish a branching time diagram which can support natural language or common sense reasoning in a satisfactory manner. However, a study of natural language reasoning suggests that such graphical representations of time should be backwards linear and forwards branching. Furthermore, the semantics of some examples from natural language reasoning on time have to be modelled in terms of a (Molinistic) TRL function. Branching time diagrams may support systems of natural language reasoning in a very useful manner. One might even consider taking ideas from durational logic into account (See [1]).

References

1. Allen, J.F., Hayes, P.J.: Moments and points in an interval-based temporal logic. Comput. Intell. **5**, 225–238 (1990)
2. Belnap, N., Green, M.: Indeterminism and the thin red line. Philos. Perspect. **8**, 365–388 (1994). Logic and Language
3. Belnap, N., Perloff, M., Xu, M.: Facing the Future: Agents and Choices in Our Indeterminist World. Oxford University Press, Oxford (2001)
4. Borghini, A., Torrengo, G.: The Metaphysics of the thin red line. In: Correia, F., Iacona, A. (eds.) Around the Tree. Synthese Library, vol. 361, pp. 105–125. Springer, Heidelberg (2013)
5. Cocchiarella, N.B.: Formal Ontology and Conceptual Realism. Springer, Heidelberg (2007)
6. Kripke, S.:Letter to A.N. Prior. A.N. Prior's Nachlass, 3 September 1958. http://nachlass. prior.aau.dk
7. Liu, H., Singh, P.: Commonsense reasoning in and over natural language. In: Negoita, M.G., Howlett, R.J., Jain, L.C. (eds.) KES 2004. LNCS (LNAI), vol. 3215, pp. 293–306. Springer, Heidelberg (2004)
8. Molina, L.D.: On Divine Foreknowledge. Cornell University Press, Ithaca (1988)
9. Prior, A.N.: Time and Modality. Oxford University Press, Oxford (1957)
10. Prior, A. N.: Letter to Saul Kripke. A.N. Prior's Nachlass, 27 October 1958, http://nachlass. prior.aau.dk

11. Prior, A.N.: Past, Present and Future. Clarendon Press, Oxford (1967)
12. Ploug, T., Øhrstrøm, P.: Branching Time, Indeterminism and Tense Logic – unveiling the Prior-Kripke letters. Synthese **188**(Nr. 3), 367–379 (2012)
13. Smart, J.J.C.: The River of time. Mind **58**(232), 483–494 (1949)
14. Smart to Prior: The Prior Collection, Box 3, The Bodleian Library, Oxford, 19 June 1955
15. Øhrstrøm, P., Hasle, P.F.: Temporal Logic: From Ancient Ideas to Artificial Intelligence. Kluwer Academics Publishers, Dordrecht (1995)
16. Øhrstrøm, P., Schärfe, H., Ploug, T.: Branching time as a conceptual structure. In: Croitoru, M., Ferré, S., Lukose, D. (eds.) ICCS 2010. LNCS, vol. 6208, pp. 125–138. Springer, Heidelberg (2010)
17. Øhrstrøm, P., Hasle, P.: Future contingents. In: Zalta, E.N. (ed.) The Stanford Encyclopedia of Philosophy. Stanford University, Stanford (2015)

Graphs and Networks

Extending GWAPs for Building Profile Aware Associative Networks

Abdelraouf Hecham[1], Madalina Croitoru[1]([✉]), Pierre Bisquert[2], and Patrice Buche[2]

[1] GraphIK, LIRMM, University of Montpellier, Montpellier, France
croitoru@lirmm.fr
[2] GraphIK, IATE, INRA, Montpellier, France

Abstract. Associative networks have been long used as a way to provide intelligent machines with a working memory and applied in various domains such as Natural Language Processing or customer associations analysis. While giving out numerous practical advantages, existing Games With a Purpose (GWAPs) for eliciting associative networks cannot be employed in certain domains (for example in customer associations analysis) due to the lack of profile based filtering. In this paper we ask the following research question: "Does considering agents profile information when constructing an associative network by a game with a purpose allows to extract subjective information that might have been lost otherwise?". In order to answer this question we present the KAT (Knowledge AcquisiTion) game that extends upon the state of the art by considering agent profiling. We formalise the game, implement it and carry out a pilot study that validates the above mentioned research hypothesis.

1 Introduction

Associative networks have been investigated as a way of representing human memory. Starting from the PhD of Quillian 1966 [14] they have been used as a way to provide intelligent machines with a working memory. A lot of work has (understandably) focused on the logical underpinning of such networks (having eventually evolved in systems such as KL ONE [4] and later on Description Logics [1] on one hand; and Conceptual Graphs [16] on the other). One aspect that has not yet been explicitly considered when constructing s is the degree of subjectivity or objectivity of the associations elicited from the humans. Such aspect is especially important in some applications that are underpinned by associative networks [7].

Associative networks have a wide range of use, from natural language processing (NLP) [13], to consumer associations analysis [10]. Each domain of application requires associative networks with different degrees of association subjectivity or objectivity. In customers analysis [17] or cognitive models [3] associative networks are only useful if they contain subjective information (i.e. reflecting the associative networks of a specific type of agents – for example the associative network of old women for 'cat food' or for a brand of makeup etc.). On the

O. Haemmerlé et al. (Eds.): ICCS 2016, LNAI 9717, pp. 43–58, 2016.
DOI: 10.1007/978-3-319-40985-6_4

other hand, in NLP, associative networks must contain objective information and therefore are being created by aggregating associative networks of different agents without any pre-condition on those agents.

In order to elicit associative networks from humans, different knowledge acquisition and elicitation techniques have been used [24]. Such techniques range from direct interviews and questionnaires to games with a purpose (GWAPs). The main idea in GWAPs is to integrate tasks (such as image tagging, video annotation, knowledge acquisition etc.) into games [20]. This technique is cheaper to implement than other knowledge acquisition methods because it relies on entertainment rather than material compensation while at the same time yielding similar or better results as shown in [19].

There are two main problems with existing GWAPs for associative network acquisition that drastically limit their use. First, unlike questionnaires, existing GWAPs do not take into account any information regarding the agents themselves. Therefore GWAPs are inherently designed to extract objective information while at the same time dismissing subjective ones. In domains such as consumer analysis such loss of information is critical. A second problem is the fact that GWAPs only elicit associative networks for concepts that come from a predefined list (dictionaries, etc.) not modifiable by external agents. This poses problems since the game itself is not easily configurable to focus on particular domains (e.g. a specific brand or product etc.) or on particular kinds of complex concepts (e.g. 'Cat' vs 'Cat owner' vs 'Good cat owner').

Therefore, in this paper, we extend upon the state of the art of GWAPs for associative network acquisition and ask the following research question: "Does considering agents profile information when constructing an associative network by a game with a purpose allows to extract subjective information that might have been lost otherwise?". In order to answer this question we present the KAT[1] (Knowledge AcquisiTion) game. We formalize the game by instantiating the formal model introduced by [6]. We implemented the game and have carried out an experimentation that validates the above mentioned research hypothesis.

The paper is structured as follows. In Sect. 3 we formally define the KAT game and highlight its properties. In Sect. 4 we describe the experimental validation of our research hypothesis using KAT Game. Finally in Sect. 5 we conclude the paper. Let us start by putting our work in the context of existing approaches in the next section.

2   Related Work and Motivating Example

Questionnaires & interviews for associative networks acquisition are limited by the resources required for acquiring large numbers of responses. The fun nature of GWAPs provide an intrinsic motivation for more participants to play and keep playing, thus, gaining greater computing power, increasing the quality of the results and lowering the cost of associative network construction [19].

---

[1] KAT Website: lirmm.cloudapp.net.

GWAPs for knowledge acquisition can be partitioned in two classes of systems [5]. On one hand GWAPs can elicit semantically structured knowledge (OntoPronto [15], GuessWhat [12], SpotTheLink [18] etc.). On the other hand, certain GWAPs elicit knowledge but in a less structured manner (WikiGame, Wikispeedia [23], Verbosity [22] or JeuxDesMots [11]). Our work is placed within the second class of GWAPs.

Existing GWAP are primarily designed to construct associative networks for NLP (namely 'Words associations') [20]. They do not take into account any information regarding the agents themselves (i.e. their profile). Therefore, the resulting associative network is an aggregated representation of what all agents think and might not be representative of a specific subset of agents due to the noise introduced by others. In domains such as consumer analysis such loss of information is critical. Therefore, in this paper we are interested in whether considering the profile of agents will impact the aggregated associative network. A second problem with existing GWAPs for knowledge acquisition is that they are not configurable, meaning that the concepts used as input to the game come from a predefined list. This rigidity when wanting to focus on certain concepts that are not commonly available.

For example, if a company wanted to know what customers associate 'Good pasta quality' with, they would not be able to use existing Knowledge acquisition GWAPs. The first reason is that there is no such complex concept in their networks. While we can find words like 'pasta', 'good' or 'quality', the concept of 'Good pasta quality' is not represented. KAT addresses this problem by allowing a user to define a domain and its concepts. Therefore the fictitious company mentioned above could input 'Good pasta quality' in the game.

Second, no profile information is considered when constructing the final associative network. Therefore potentially meaningful associations are drowned by the noise of the large quantity of unfiltered associations (as demonstrated by our experiment in Sect. 4). KAT handles this issue by allowing the user to define profile criteria that will generate various profiles (either explicitly or implicitly through profile similarity degree), these profiles will allow the filtering of associations corresponding to classes of customers the company wants to specifically target.

The associative network elicited by the GWAP presented in this paper is a graph. The nodes represent labeled concepts and the edges associations between these objects. Each player is asked a set of questions that will determine his profile (for example gender, interests, qualifications etc.). The ordered list of associations given by players with similar profiles will be aggregated in order to build the final associative network. The players are also asked if the associated concepts hold a positive, negative or neutral connotation. With respect to the state of the art, the associative network elicited by KAT has the following novel features:

- The initial game concepts and profile questions are dynamically created by players.
- The associations between concepts are computed by taking into account the profile of the players.
- The associated concepts are ordered.
- The associated concepts hold a positive, negative or neutral connotation.

In the next section we formalise the KAT game and explore its properties.

3 KAT: Formalisation and Properties

In this paper we propose a method to build an associative network using a GWAP that takes into account the profile of the players. Here, the associations creation task is transformed into elements of a game where profile teamed players construct and validate associations as a consequence of playing the game, rather than by performing a more traditional direct questions-answering task. In order to explain and define formally the associative network we build, several concepts need to be introduced first: concept, opinion, domain, profile and association.

3.1 Preliminary Notions

A concept is a well formed formula of $\mathcal{L}$, where $\mathcal{L}$ is a propositional logic language based on a set of propositional symbols $\mathcal{V}$.

Given a set of concepts $C \subseteq \mathcal{L}$, an opinion is a function that maps a concept $\varphi \in C$ to a set of appreciations in $\mathcal{L_O}$, where $\mathcal{L_O}$ is an opinion representation language.

Opinions can be represented using different languages. In this paper, we will use $\mathcal{L_O}$ that defines three levels of opinions {like, dislike, indifferent} denoted $\{\oplus, \ominus, \odot\}$ respectively.

An associative network is a binary relation $\mathcal{AN} \subseteq (\mathcal{L} \cup \mathcal{L_O}) \times (\mathcal{L} \cup \mathcal{L_O})$, that can associate a concept with an opinion, a concept with a concept or an opinion with an opinion. The $\mathcal{AN}$ creation is the task of eliciting from an agent the associations between concepts and possibly his opinions about them. Then, by aggregation, we can construct the associative network of a certain profile (group of agents).

The main novel aspect introduced in the game by this paper is the notion of profile. A profile defines the various classes of game players. Given $\mathcal{P}$ the set of agents (participants), a profile Pr_j is a subset of $\mathcal{P}$. There are many ways of building a profile, defined explicitly by fixing a set of criteria that have to be met (e.g. participants that are women and are at least 30 years old), or defined implicitly using a 'similarity degree' based on a set of weighted criteria (e.g. participants that are at least 30 % similar). To determine the similarity degree regarding a set of weighted criteria we use *OKCupid* matching algorithm [8]. These criteria are encapsulated with a concept set in a domain.

Definition 1 (Domain). *A domain is a tuple* $\mathcal{D} = (\mathcal{C}^{\mathcal{D}}, \mathcal{CR}^{\mathcal{D}})$ *s.t:*

- $C^{\mathcal{D}}$ is a set of concepts.
- $C\mathcal{R}^{\mathcal{D}} = \{(C\mathcal{R}_1, importance_1), ..., (C\mathcal{R}_n, importance_n)\}$ is a set of profile criteria weighted by importance s.t:
  - $C\mathcal{R}_n$ is a profile criterion (e.g. age, gender, expertise...etc.).
  - $importance \in \{notAtAllImportant, aLittleImportant, somewhatImportant, veryImportant, mandatory\}$.

The $\mathcal{AN}$ of an agent can be constructed by asking multiple agents of the same profile three simple questions for every key concept $\varphi \in C$ of the domain:

1. What do you associate this concept φ with?
2. Classify these associations from most relevant to least relevant.
3. Give -if possible- an opinion for every one of these related concepts.

The construction of the $\mathcal{AN}$ of each profile related to the domain of study by a game with a purpose provides an intrinsic motivation for players (agents) to keep playing, which can increase the quality of the associative networks created. By playing the KAT game, the $\mathcal{AN}$ of a profile would then be constructed by aggregating the associative networks of all the agents in that profile.

The output of the KAT Game is the profile associative network for the given concept, and defined as the aggregation by profile of the agents associative networks.

Definition 2 (Agent Associative network). *An associative network of an agent p_i is a tuple $AN_{p_i} = (R, op, weight)$ where*

- $R \subseteq C \times \mathcal{L}$ is a binary relation called association that maps a concept to another wff of $\mathcal{L}$.
- $op : \mathcal{L} \to \mathcal{L}_O$ is a function that maps a concept to an opinion in $\mathcal{L}_O$.
- $weight : R \to \mathbb{N}$ is a function that maps each association $r \in R$ to an integer called weight of r.

The profile associative network is defined as follows:

Definition 3 (Profile Associative Network). *An associative network AN_{Pr_j} for a profile $Pr_j \subseteq \mathcal{P}$ is the aggregation of the associative networks of all agents $p_i \in Pr_j$: $AN_{Pr_j} = \biguplus_{i=1}^{n} AN_{p_i}$ where $\biguplus$ is an aggregation function s.t: $\biguplus_{i=1}^{n} AN_{p_i} = (\biguplus_{i=1}^{n} R_i, \biguplus_{i=1}^{n} op_i, \biguplus_{i=1}^{n} weight_i)$.*

AN_{Pr_j} is obtained by computing the aggregation of the relations which is the union of all relations $\biguplus_{i=1}^{n} R_i = \bigcup_{i=1}^{n} R_i$, the aggregation of opinions which is the median of the set of all opinions $\biguplus_{i=1}^{n} op_i = median\{op\}$, and the aggregation of weights which is the arithmetic mean of all weights divided by the number of occurrences $\biguplus_{i=1}^{n} weight_i = \frac{1}{n^2} \sum_{i=1}^{n} weight$.

3.2   Game Description

In [21] three game-structure templates are defined: *output-agreement games, inversion-problem games, and input-agreement games*. KAT falls into the class of output-agreement games, these games have the following structure:

- **Initial setup:** Two strangers are randomly chosen by the game itself from among all potential players;
- **Rules:** In each round, both are given the same input. Players cannot see each others outputs or communicate with one another;
- **Winning condition:** Both players must produce the same output; they do not have to produce it at the same time but must produce it at some point while the input is displayed on screen.

KAT is a two player blind game based on output-agreement. Blind implies that the other player is unknown until the end. Output-agreement means that both players will always score the same amount of points according to what they have proposed in common. The game is designed to inherently ensure both data correctness and produce a representation as close as possible to the typical associative network of a defined profile. KAT is a cooperative game in which two players strive to gain points by entering terms (concepts) they associate with the input, ordering them, and by giving their opinions about these concepts. (e.g. in Fig. 1, the player is playing the 'Cat Owner' concept that he associates with 'Woman', 'Cat lover', 'Dog hater' and 'House owner' in this order, the player then gives his opinion about some of his associations, he likes 'Woman', dislikes 'Dog hater' and is indifferent towards 'house owner').

Before the beginning of a game session, players select a domain, then they proceed to answer profile related questions, these answers will be taken into account in order to pair players with similar profile against each other. (As shown in Fig. 2, the player answers a set of questions defined by the creator of the domain 'Cat' such as their gender, the fact whether they have allergies to cat or not etc.).

At the beginning of a game session two players are paired. They are shown an input (a concept) that was randomly chosen from the set of concepts associated with the played domain. Each player can enter up to 10 ordered terms that are related to the input, then they can give their opinion regarding each one of these terms. Without seeing or communicating with each other, the players must agree on as many terms as possible, they gain more points if they agree on the order of the terms and their opinion. By agreeing with the partner on terms and their order for as many concepts as possible we ensure that the players associations are accurate representation of the associative network of a typical agent from that profile. While KAT relies on comparison between two players inputs to determine the score and the winning condition, both players are not necessarily playing at the same time. To implement this we follow [19] where a game session is randomly chosen between two types: either the player will complete a game previously created by another player (concluding a game) where his inputs will be compared to a previously stored one from the other player or start a new

Fig. 1. Example of a game session for the concept 'Cat owner'. The player inputs his/her associations, orders them, and optionally gives his/her opinions.

Fig. 2. Pre-game profile selection. Before playing the concepts of a domain, a player has to answer some questions in order to define his/her profile.

game by storing his input for further comparison. When concluding a game, the player is given his score immediately, and the other player will be notified of his score.

3.3 Game Formalisation

In order to formally define our KAT game, we use the GWAP formal model introduced by [8], this model was defined to be generic enough to incorporate all types of GWAP games (input-agreement, output-agreement and inversion-problem games [20]). For the sake of simplicity, we will only define components that are necessary to formalize an output-agreement GWAP. In KAT game, an input data $\mathfrak{D}$ is a concept $\varphi \in \mathcal{C}$ that is composed of a name (c_{name}), a description ($c_{description}$) and potentially an image (c_{image}). A concept has only one attribute that is a set of associations $Attr = (associations)$.

Definition 4 (The Knowledge AssociaTion Game). *KAT is a 4-tuple* ($\mathcal{SGPD}$, $\mathcal{GR}$, $\mathcal{GF}$, $\mathcal{ANS}$) *where sets:*

1. $\mathcal{SGPD} = (e, \mathcal{F}, \mathcal{G})$ *is the Social Game Problem Domain:*
   (A) *e is the problem we want to solve which is to collect the associations metadata of an input data $\mathfrak{D}$.*
   (B) $\mathcal{F} = \{f_i | i = 1, ..., y\}$ *is the answers domain. (Solutions to the problem e, where f_i is an association, can only be in $\mathcal{F}$).*
   (C) $\mathcal{G} : e \times \mathcal{F} \to \mathfrak{R} \in [0..1]$ *is a function that determine whether an answer is correct to a problem.*
2. $\mathcal{GR} = (\mathcal{P}, \mathcal{I}, \mathcal{O}, \mathcal{W})$ *represents the rules of the social game*
   (a) $\mathcal{P} = \{p_1, p_2\}$ *is the set of players.*
   (b) $\mathcal{I}$ *is the set of input given to the player for solving the problem of input $\mathfrak{D}$ during a game:* $I(p_1) = I(p_2) = \{c_{name}, c_{description}, c_{image}\}$.
   (c) $\mathcal{O}(p_j) = \{o_m^j | m = 1, ..., n\}$ *is the set of outputs provided by the player p_j for solving the problem of input $\mathfrak{D}$ during the game.*
   (d) $W(p_j)$ *is the reward that the player can receive for solving the problem of input $\mathfrak{D}$ during a game where* $W(p_j) \in \{w_i | i = 1, ..., y\}$. *Players will receive a reward when achieving the winning condition of the game.*
3. $GF = \{pSel, eSel, tMax, pNum, GM\}$ *represents the flow of a GWAP.*
   (a) $pSel()$ *is a procedure that selects players and assigns roles to them.*
   (b) $eSel()$ *is a procedure that picks a problem from the problem set.*
   (c) $tMax$ *is the maximum duration of a game.*
   (d) $pNum = 2$ *is the number of players of a game.*
   (e) $GM \in \{collaborative, competitive, hybrid\}$ *is the mechanism of the game. KAT is a collaborative game.*
4. $\mathcal{ANS} = (\xi, \tau)$ *represents answer extraction. It defines how answers are generated for a problem based on all the games played. ξ is a data structure and τ is a frequency threshold for accepting an answer. if an answers is given by at least τ players it will be added to the set of correct answers $c\mathcal{F} \subseteq \mathcal{F}$.*

The objective of a KAT game session is to solve the problem e of collecting associations for a given concept φ, these associations have to be within the domain correct answers $c\mathcal{F}$. The problem e is selected using the $eSel()$ procedure that randomly picks a concept from the set of concepts $C^{\mathcal{D}}$ of the chosen domain $\mathcal{D}$.

In each session, two-player (p_1 and p_2) aim to provide a common output (($\mathcal{O}(p_1) \cap c\mathcal{F}) \cap (\mathcal{O}(p_2) \cap c\mathcal{F}) \neq \varnothing$). These players are selected to compete against each other by the $pSel()$ procedure that takes into account the degree of similarity between players and attempts to pit against each other players with similar profiles.

KAT is a collaborative game, both players are given the same input $I(p_1) = I(p_2)$, they try to fulfill their assigned tasks by helping each other complete the task and achieve the winning condition, they receive rewards $W(p_1)$ and $W(p_2)$ (called score) that depend on the number of common correct answers $|(\mathcal{O}(p_1) \cap c\mathcal{F}) \cap (\mathcal{O}(p_2) \cap c\mathcal{F})|$, the number of common opinions, the order of the given outputs, and the degree of similarity between the two players profile.

3.4 KAT Game Properties

One of the main properties of a GWAP is the **type of information** outputted by the game. Such information can be objective or subjective. Subjective information is information that is related to the mental model of one particular group of agents. Such information is only shared between a specific type of agents (for example, association such as (cat, child) is more recurrent between 'old women' whereas (cat, useless) is more frequently given by 'young men'). Objective information on the other hand is not affected by users profile, it is shared by almost all agents. An example of objective information could be the association (cat, animal).

For a given problem e, the probability that players' outputs are accepted depends on $|\mathcal{O}(p_1) \cap \mathcal{O}(p_2)|$, where $\mathcal{O}(p_1) \cap \mathcal{O}(p_2)$ is the set of potential outputs shared by players. Objective informations -by definition- have higher chance of being shared by various players, thus, the correct common output between player 1 and player 2 is almost the same as the correct output of player 1 together with the correct output of player 2: $(\mathcal{O}(p_1) \cap O(p_2) \cap c\mathcal{F}) \approx ((\mathcal{O}(p_1) \cap c\mathcal{F}) \cup (\mathcal{O}(p_2) \cap c\mathcal{F}))$. Similarly, subjective informations have lower probability of being shared by different agents, therefore, the correct common output of player 1 and player 2 is much lower than the correct output of player 1 together with the correct output of player 2: $(\mathcal{O}(p_1) \cap O(p_2) \cap c\mathcal{F}) \ll ((\mathcal{O}(p_1) \cap c\mathcal{F}) \cup (\mathcal{O}(p_2) \cap c\mathcal{F}))$.

Existing associations acquisition GWAPs extract truly objective informations that are shared between the majority of agents, and in the same process, they inherently eliminate subjective informations. Whereas, by using profiles in KAT, we can 'isolate' specific groups of agents, and within these groups, a previously subjective information might become an objective one (e.g. the association (cat, child) is subjective if we consider all agents and thus it won't be included in the final associative network for 'cat', but if we only consider 'old women', this association might become objective because it is shared by most considered agents). To say this otherwise, subjective information can become objective when restricted to a profile.

While having the ability to define the wanted profile explicitly allows a KAT user to eliminate 'noise' information from the resulting associative network, having the ability to implicitly define profiles using a 'profile similarity degree' allows KAT to smoothly transition from truly objective information (similarity degree 0 %) to subjectively objective information (similarity degree 100 %) i.e. the higher the similarity degree the better the chance is to extract subjective information.

4   KAT Experimental Validation

In this section we present how we conducted the experimental validation of
the research hypothesis presented in the paper. We recall the reader that the
experiment aims to validate the following research question: "Does considering
agents profile information when constructing an associative network by a game
with a purpose allows to extract subjective information that might have been
lost otherwise?". We have chosen to build an associative network of the Durum
Wheat domain. The first experimentation was carried out with the experts of
the DUR-DUR project, a French government funded project aiming at studying
the sustainability of the Durum Wheat food chain. The second experimentation
was carried out with undergraduate students in computer science in a life long
learning class.

4.1   Experiment Design

We are interested in the impact of taking into account participants profiles when
generating the aggregated set of associations. We will consider different profile
similarity degrees and generate the corresponding associative networks. We then
evaluate how representative these associative networks are. Our study requires
multiple game sessions to be played for a specific set of concepts by participants
with different profiles. We quantify the impact of a participant profile on the
aggregated set of associations by using a relevance metric. We define the relevance of an association regarding a set of aggregated associations as how many
participants agreed on it compared to the total number of participants taken
into account in the aggregated associations.

Definition 5 (Relevance of an association). *Let $AN_{Pr_j} = \biguplus_{i=1}^{n} AN_{p_i}$ be
the aggregated associative network of a set of players Pr_j for the concept $\varphi \in \mathcal{C}$
where AN_{p_i} is the associative network of a player $p_i \in Pr_j$ for φ. We define the
relevance of an association relation $r \in R$ of AN_{p_i} regarding AN_{Pr_j} as:*

$$relevance(r, AN_{Pr_j}) = \frac{occurence(r, AN_{Pr_j})}{|Pr_j|} \text{ where } occurence(r, AN_{Pr_j}) \text{ is the}$$

number of participants that had r in their Association Network for φ.

We define the relevance of an associative network AN_{p_i} as the average relevance of all its associations. To compare two aggregations of associative networks
we simply need to compare their relevance for each of the participants. The
higher the average relevance the more representative the aggregated associative
network is. Please note that the more representative the associative network the
better subjective associations are exposed. This is due to the fact that subjective
information has a lower probability of being shared by many players, thus by
definition, a subjective association will have a lower relevance when we consider
all players.

4.2 Profile Criteria

We chose to conduct our experiments in the domain of the DUR-DUR project aiming at studying the Durum Wheat food chain. Our participants got shown 15 concepts specific to the DUR-DUR domain. We built profiles implicitly using two criteria Cr_1 and Cr_2.

Cr_1 states whether the participant is part of the DUR-DUR project or not (possible answers are "Yes" or "No"), we gave this criterion an importance of *veryImportant* because we judge that being an expert or not in this domain will greatly affect what the concepts of that domain are associated with. Cr_2 determines the Area of expertise, (possible answers are "Agronomy", "Transformation", "Socio-economic", "LCA (Life Cycle Analysis)", "Computer Science", "None of the above"). We gave this criterion an importance of *somewhatImportant* based on the fact that it is significant but not as critical as Cr_1.

Please keep in mind that our hypothesis states that associative networks vary depending on players profiles. The criteria and the corresponding answers allow to partition the players into different profiles and can be defined as needed by the game designer.

4.3 Experiment Execution

The experiment was implemented in two parts. A first experimentation was carried out with 9 experts of the DUR-DUR project (7 of them were Transformation experts while the rest (2) were LCA experts). A second experimentation was carried out with 15 undergraduate students in computer science from the University Institute of Technology (IUT) in Montpelier. All 24 participants played 15 game sessions successfully completed. The experiment took approximately 30 min per participant. The experiment was carried out in three phases. The first phase was initial training, all participants were shown how to play and a few game sessions were played for testing purposes. The second phase was when the experiment itself was carried out. Finally, the third phase was data processing where some associations given in French were manually translated in English in order to be taken into account in the statistical analysis.

Concretely, the participants played a game session where they were shown the concept name in English, a description in French and optionally a photo representing the concept, (e.g. For 'Pasta Quality', the game displayed 'Pasta Quality' as concept name, 'Qualité des pates' as a description, and an image of pasta). The participant was then asked to give associations in English, organize them from most relevant to least relevant and, optionally, give an appreciation for each association.

4.4 Statistical Analysis

The participants played 15 game sessions giving a total of 1623 associations. For lack of space, we will only present 3 concepts, of which we show the full results for the concepts 'Pasta Quality' and the partial results for 'Protein content' and 'Couscous Processing' (shown in Tables 1, 2 and 3. Please note that associations are ordered from most relevant on top to least relevant at the end of the tables).

To establish if there was an overall effect we computed the average relevance for each profile partition obtained using a different similarity degree. We can see that higher similarity degree yields higher relevance meaning that $(\mathcal{O}(p_i) \cap \mathcal{O}(p_j))$ is getting closer to $(\mathcal{O}(p_i) \cup \mathcal{O}(p_j))$, thus more subjective associations are being exposed. For example, in Table 1 we can notice that for similarity degree superior to 0 % the average relevance of 0.22 is greater than 0.08 that corresponds to a similarity degree of 0 %.

Please note that by using a similarity degree of 0 % (and therefore not considering the profile information) the associations elicited are highly objective. This might pose a problem for certain applications as the associations cannot be used. For instance please look at the associations of *Pasta Quality* that, in terms of objective associations, yields *Italy* and *Cooking Time* as the first two associations. Similarly *Protein Content* yields *Meat* and *Muscle* and *Couscous Processing* yields *Semolina* and *Arab*. These associations might be useful for NLP techniques. However, if pursuing an analysis of expert associations their subjective information gets completely drowned in the noise of all data. Furthermore, opinions for some associations might change depending on the similarity degree, for example, in the associative networks for *Pasta Quality*, experts think positively of *Gluten* while non-experts are totally indifferent.

While this is obvious in all the three examples we chose to show in this paper, let us closely analyse *Couscous Processing*. By using a similarity degree different from 0 % we can expose valuable subjective associations of the experts (e.g. *Semolina* and *Rolling* and *Durum Wheat*) that have been drowned in the objective associations of everybody (like *Arab*, *Tajine*). Also please note that the subjective associations of IUT students are similar to the objective associations of everybody due to the higher number of students. This clearly shows how not making the distinction objective/subjective can suppress the associations of under-represented types of agents.

4.5   Threats to Validity

The sources of threats to validity to our work are either based on flaws related to the experimental setting of our validation or the statistical significance of our results. Regarding the last aspect, standard statistical tests (such as the test of Pearson [9]) cannot be employed as the whole alphabet of correlations is not known in advance. All one can do to alleviate such shortcoming is to increase the dataset (of both experts and non experts). While finding non experts is not difficult, finding experts is a much more challenging and time consuming task. We plan to carry out a longer experiment to include more experts in future work but for the moment this paper only presents a pilot study.

Regarding the flaws related to the experimental setting we distinguish two kinds of potential threats to validity. First we need to be aware of a potential bias related to the way participants will adapt in order to increase the number

Table 1. Associative networks for 'Pasta Quality'

Similarity degree of 0 %		Similarity degree ∈]0; 83.36] %			
No criteria		Experts		Non-experts	
Italy	⊕	Yellowness	⊕	Italy	⊕
Cooking time	⊙	Color	⊙	Cooking Time	⊙
Taste	⊙	Protein Content	⊕	Price	⊙
Protein Content	⊕	Texture	⊕	Taste	⊙
Yellowness	⊕	Stickiness	⊕	Brand	⊙
Nutrition	⊕	Starch	⊙	Nutrition	⊕
Price	⊙	Cooking loss	⊖	Slow sugar	⊕
Color	⊙	Taste	⊙	Gluten	⊙
Gluten	⊕	Drying temperature	⊕	Tomato sauce	⊕
Brand	⊙	Hydration	⊕	Panzanni	⊕
Average relevance: 0.08		Average relevance: 0.22			

Similarity degree ∈]83.36; 100] %					
LCA experts		Transformation experts		Non-experts	
Yellowness	⊕	Color	⊙	Italy	⊕
Texture	⊕	Protein Content	⊕	Cooking Time	⊙
Strach	⊕	Yellowness	⊕	Price	⊙
Nutrition	⊕	Stickiness	⊕	Taste	⊙
Protein nature	⊙	Drying temperature	⊕	Brand	⊙
Network	⊙	Overcooking resistance	⊕	Nutrition	⊕
Cropping system	⊕	Gluten	⊕	Slow sugar	⊕
Quantity	⊕	Cooking loss	⊖	Gluten	⊙
Brightness	⊕	Texture	⊕	Tomato Sauce	⊕
Color imperfection	⊖	Viscoelasticity	⊕	Panzanni	⊕

Average relevance: 0.38

Table 2. Associative networks for 'Protein Content'

Similarity degree of 0 %		Similarity degree ∈]0; 83.36] %			
No criteria		Experts		Non-experts	
Meat	⊕	Quality	⊕	Meat	⊕
Muscle	⊕	Gluten	⊕	Muscle	⊕
Nutrition	⊕	Gliadine	⊕	Eggs	⊕
Quality	⊕	Nutrition	⊕	Nutrition	⊕
Eggs	⊕	Network	⊕	Chicken	⊕
...	...	...	...	...	...
Average relevance: 0.04		Average relevance: 0.23			

Table 3. Associative networks for 'Couscous Processing'

Similarity degree of 0 %		Similarity degree $\in$]0; 83.36] %			
No criteria		Experts		Non-experts	
Semolina	$\oplus$	Semolina	$\oplus$	Arab	$\odot$
Arab	$\odot$	Rolling	$\oplus$	Tajine	$\oplus$
Water	$\oplus$	Durum Wheat	$\odot$	Semolina	$\odot$
Tajine	$\oplus$	Agglomeration	$\oplus$	Water	$\oplus$
Rolling	$\oplus$	Water	$\oplus$	Maghreb	$\odot$
...	...	...	...	...	...
Average relevance: 0.04		Average relevance: 0.19			

of points they score (and thus think of words closer to what the others would write as opposed to their own associations). Please note though that this bias is something that we want as it reinforces the way the aggregated network is constructed. Second, cheating is always possible when the game is conducted with participants that share the same computer room as they could purposely ask the others what their associations are. To the best of our knowledge such conduct has not occurred in any of the game sessions.

5   Conclusion

This paper extended upon the state of the art and presented a GWAP that allows to elicit associative networks and expose subjective associations. We formally defined the game and carried out a pilot study in order to analyse its behavior on a real world scenario. The experimental pilot study confirmed the research hypothesis of the paper and namely that KAT can successfully extract the subjective information that might have been lost within objective informations otherwise.

The future work directions this work opens are numerous. First the scoring aspect of the game has yet not been examined in detail. Improving upon this aspect might increase the hedonistic value of the game. Second we can better expose the elicited data by providing dedicated tools and interfaces. Last, please note that while the construction of the associative network of each profile related to the domain of study might be accomplished using text mining techniques, such elements could be incorporated in the game to enrich its basic set of concepts.

We plan to use the KAT game to elicit the association table described in the paper of [2,3]. However this work could also be used in reverse profile elicitation where based on the various associations of an agent, one can define its profile. Therefore such work could also be relevant in the context of recommender systems or customer analysis.

Acknowledgements. The authors acknowledge the support of ANS1 1208 IATE INCOM INRA grant, ANR grants ASPIQ (ANR-12- BS02-0003), QUALINCA (ANR-12-0012) and DUR-DUR (ANR-13-ALID-0002). The work of the second author has been carried out part of the research delegation at INRA MISTEA Montpellier and INRA IATE CEPIA Axe 5 Montpellier. The authors are grateful to DUR-DUR participants and IUT AS students for the help with the experimentation.

References

1. Baader, F.: The Description Logic Handbook: Theory, Implementation and Applications. Cambridge University Press, Cambridge (2003)
2. Bisquert, P., Croitoru, M., Dupin de Saint-Cyr, F.: Four ways to evaluate arguments according to agent engagement. In: Guo, Y., Friston, K., Aldo, F., Hill, S., Peng, H. (eds.) BIH 2015. LNCS, vol. 9250, pp. 445–456. Springer, Heidelberg (2015)
3. Bisquert, P., Croitoru, M., de Saint-Cyr, F.D.: Towards a dual process cognitive model for argument evaluation. In: Beierle, C., Dekhtyar, A. (eds.) SUM 2015. LNCS, vol. 9310, pp. 298–313. Springer, Heidelberg (2015)
4. Brachman, R.J., Schmolze, J.G.: An overview of the KL-ONE knowledge representation system. Cogn. Sci. **9**(2), 171–216 (1985)
5. Cambria, E., Rajagopal, D., Kwok, K., Sepulveda, J.: Gecka: game engine for commonsense knowledge acquisition. In: The Twenty-Eighth International Flairs Conference (2015)
6. Chan, K.T., King, I., Yuen, M.-C.: Mathematical modeling of social games. In: International Conference on Computational Science and Engineering, CSE 2009, vol. 4, pp. 1205–1210. IEEE (2009)
7. Findler, N.V.: Associative Networks: Representation and Use of Knowledge by Computers. Academic Press, New York (2014)
8. Fry, H.: The Mathematics of Love: Patterns, Proofs, and the Search for the Ultimate Equation. Simon and Schuster, New York (2015)
9. Greenwood, P.E., Nikulin, M.S.: A Guide to Chi-Squared Testing, vol. 280. Wiley, New York (1996)
10. Henderson, G.R., Iacobucci, D., Calder, B.J.: Brand diagnostics: mapping branding effects using consumer associative networks. Eur. J. Oper. Res. **111**(2), 306–327 (1998)
11. Lafourcade, M.: Making people play for lexical acquisition with the JeuxDeMots prototype. In: 7th International Symposium on Natural Language Processing, SNLP 2007, p. 7 (2007)
12. Markotschi, T., Völker, J.: Guesswhat?!-Human intelligence for mining linked data (2010)
13. Miller, G.A.: Wordnet: a lexical database for english. Commun. ACM **38**(11), 39–41 (1995)
14. Quillan, M.R.: Semantic memory. Technical report, DTIC Document (1966)
15. Siorpaes, K., Hepp, M.: Games with a purpose for the semantic web. IEEE Intell. Syst. **3**, 50–60 (2008)
16. Sowa, J.F.: Conceptual graphs for a data base interface. IBM J. Res. Dev. **20**(4), 336–357 (1976)
17. Teichert, T.A., Schöntag, K.: Exploring consumer knowledge structures using associative network analysis. Psychol. Mark. **27**(4), 369–398 (2010)

18. Thaler, S., Simperl, E.P.B., Siorpaes, K.: SpotTheLink: a game for ontology alignment. Wissensmanagement **182**, 246–253 (2011)
19. Vannella, D., Jurgens, D., Scarfini, D., Toscani, D., Navigli, R.: Validating and extending semantic knowledge bases using video games with a purpose. ACL **1**, 1294–1304 (2014)
20. Von Ahn, L.: Games with a purpose. Computer **39**(6), 92–94 (2006)
21. Von Ahn, L., Dabbish, L.: Designing games with a purpose. Commun. ACM **51**(8), 58–67 (2008)
22. Von Ahn, L., Kedia, M., Blum, M.: Verbosity: a game for collecting common-sense facts. In: Proceedings of the SIGCHI Conference on Human Factors in Computing Systems, pp. 75–78. ACM (2006)
23. West, R., Pineau, J., Precup, D.: Wikispeedia: an online game for inferring semantic distances between concepts. In: IJCAI, pp. 1598–1603 (2009)
24. Wilson, J.R., Sharples, S.: Evaluation of Human Work. CRC Press, London (2015)

Representing Multi-scale Datalog+/− Using Hierarchical Graphs

Cornelius Croitoru[1,2] and Madalina Croitoru[1,2(✉)]

[1] Faculty of Computer Science, Al. I. Cuza University, Iaşi, Romania
[2] GraphIK, University of Montpellier, Montpellier, France
croitoru@lirmm.fr

Abstract. We introduce a multi scale knowledge representation and reasoning formalism of Datalog+/− knowledge bases. This is defined on a novel graph transformation system that highlights a new type of rendering based on the additional expansion of relation nodes. Querying and integration capabilities of our approach are based on a FOL sound and complete homomorphism.

1 Introduction

The set of requirements for knowledge representation formalisms must include *(i)* the existence of a declarative semantics, *(ii)* a logical foundation, and *(iii)* the possibility of representing structured knowledge [2]. While many languages have followed these three directions, a lot of existing work focused mainly on the first two aspects. Here we address the third requirement and namely the need to represent hierarchical, multi scale knowledge. Therefore, our representation structures fulfil all these three conditions in a formal way. By hierarchical, multi scale, knowledge we understand knowledge that can be represented at different level of granularity. For instance, we can see the human body as made out of body parts such as hands, legs, lungs etc. or we can zoom in and look at the muscles and the bones or we can further zoom in and see how minerals and organic substances interact in our body. Such levels are not disconnected - a lack of Mg in the body can lead to muscle spasms that can lead to tingling in the legs. Multi scale knowledge bases are commonly used in Life Sciences [5] but not only. Supply Chain Management [16], Information Integration [19], Sensor Networks [1], Policy Rules [13] etc. all require to represent and reason about knowledge at various levels of granularities while being able to go from one level to the other easily.

The representation we propose is using the notion of a transitional description. The transitional description allows to go from one level of granularity to the next. This mechanism is used to define inductively hierarchical structures of depth d. We build upon the state of the art and, in this paper, consider the Datalog+/− language. By considering n-ary predicates and existential rules (i.e. rules that allow for existentially quantified variables in the conclusion) this language generalises certain Description Logics. Even if not endowed with a

© Springer International Publishing Switzerland 2016
O. Haemmerlé et al. (Eds.): ICCS 2016, LNAI 9717, pp. 59–71, 2016.
DOI: 10.1007/978-3-319-40985-6_5

graphical depiction, Datalog+/− is logically equivalent to Conceptual Graphs with Rules and Negative Constraints. In this paper, while considering the core logical language of Datalog+/− we endow it with graph based logically sound and complete semantics.

The paper is structured as follows. In Sect. 2 we explain the choice of logical language and place ourselves within the state of the art for representing hierarchical knowledge. In Sect. 3 we recall basic notions needed throughout the paper such as facts, rules, knowledge base, etc. We also show how to endow the Datalog+/− language with a graph based syntax while staying sound and complete wrt semantics. Section 4 presents the hierarchical knowledge representation and reasoning formalism and Sect. 5 concludes the paper.

2   State of the Art

In this paper we consider a rule based language that gains more and more interest from a practical point of view, Datalog+/− [7]. We consider existential variables in the head of the rules as well as n-ary predicates and conflicts (and generalise certain subsets of Description Logics (e.g. DL-Lite) [3,8]). The tractability conditions of the considered rule based language rely on different saturation (chase) methods [17]. The language can be equivalently seen in a logically sound and complete graph based representation [4,14].

The data structure discussed here evolved from Conceptual Graphs [18], Nested Conceptual Graphs [11] and respectively Layered Conceptual Graphs [12]. The idea of a detailed context of knowledge can be traced back to the definition of Simple Conceptual Graphs (SCGs) [18], to the work of [15] and to the definition of the more elaborate Nested Conceptual Graphs [11]. The querying capabilities associated with our approach are supported by the logically sound homomorphism operation, which is defined between a query and the hierarchical structure.

Except [12], existing work in representing hierarchical knowledge in diagrammatic way does not relate to the context in which the complex nodes appear. The complex nodes behave like *glass boxes*, corresponding to a "zoom" action. In [12] the authors use a similar notion of multi level granularity knowledge but their approach is closely following Conceptual Graphs. In this paper the language used is more generic following [7]. Last, a hierarchical extension of Datalog has also been proposed by [6] but the approach suffers from the lack of graph based rendering of the transitions between levels.

3   Basic Notions

We consider constants but no other functional symbols; a vocabulary W is composed of a set of predicates P and a set of constants C. Constants identify the individuals in the knowledge base and predicates represent n-ary relations between such individuals. We also consider X, a set of variables in the knowledge base.

Definition 1 (Vocabulary). *Let C be a set of constants and P a set of predicates. A vocabulary is a pair $W = (P, C)$ and arity is a function from P to $\mathbb{N}$. For all $p \in P, arity(p) = i$ means that the predicate p has arity i.*

We will consider an infinite set X of variables, disjoint from P and C. A term is an element of $C \cup X$. An atom is of form $p(t_1,...,t_k)$, where p is a predicate of arity k in W and the t_i are terms. For a given atom A, we note $terms(A)$, $csts(A)$ and $vars(A)$ the terms, constants and variables occurring in A.

Definition 2 (Fact). *A fact is a finite, but possibly empty, set of atoms on a vocabulary. For a given fact F, we note $atoms(F)$ the atoms occurring in F.*

Example. Let us consider a vocabulary $W = (P, C)$. $P = \{$man, woman$\}$, $C = \{Bob, Alice\}$ and $arity = \{($man, 1$), ($woman, 1$)\}$. $man(Bob)$ and $woman(Alice)$ are two distinct atoms on W, and $F = \{man(Bob), woman(Alice)\}$ a fact.

We can represent facts as labelled ordered bipartite graphs where one class of partition represents the concepts (i.e. the unary predicates) and the other the relations. Such representation is well known in the literature (see [9] or [18]).

A *bipartite graph* is a graph $G = (V_G, E_G)$ with the nodes set $V_G = V_C \cup V_R$, where V_C and V_R are finite disjoint nonempty sets, and each edge $e \in E_G$ is a two element set $e = \{v_C, v_R\}$, where $v_C \in V_C$ and $v_R \in V_R$. Usually, a bipartite graph G is denoted as $G = (V_C, V_R; E_G)$. We call $G^\emptyset$ the empty bipartite graph without nodes and edges.

Let $G = (V_C, V_R; E_G)$ be a bipartite graph. The number of edges incident to a node $v \in V(G)$ is the degree, $d_G(v)$, of the node v. If, for each $v_R \in V_R$ there is a linear order $e_1 = \{v_R, v_1\}, \ldots, e_k = \{v_R, v_k\}$ on the set of edges incident to v_R (where $k = d_g(v)$), then G is called an *ordered bipartite graph*. A simple way to express that G is ordered is to provide a labelling $l : E_G \rightarrow \{1, \ldots, |V_C|\}$ with $l(\{v_R, w\}) =$ index of the edge $\{v_R, w\}$ in the above ordering of the edges incident in G to v_R. l is called a *order labelling* of the edges of G. We denote an ordered bipartite graph by $G = (V_C, V_R; E_G, l)$.

For a vertex $v \in V_C \cup V_R$, the symbol $N_G(v)$ denotes its neighbours set, i.e. $N_G(v) = \{w \in V_C \cup V_R | \{v, w\} \in E_G\}$. Similarly, if $A \subseteq V_R \cup V_C$, the set of its neighbours is $N_G(A) = \cup_{v \in A} N_G(v) - A$. If G is an ordered bipartite graph, then for each $r \in V_R$, the symbol $N_G^i(r)$ denotes the i-th neighbour of r, i.e. $v = N_G^i(r)$ if and only if $\{r, v\} \in E_G$ and $l(\{r, v\}) = i$.

Throughout this paper we use a particular type of subgraph of a bipartite graph: $G^1 = (V_C^1, V_R^1; E_G^1)$ is a subgraph of $G = (V_C, V_R; E_G)$ if $V_C^1 \subseteq V_C, V_R^1 \subseteq V_R$, $N_G(V_R^1) \subseteq V_C^1$ and $E_G^1 = \{\{v, w\} \in E_G | v \in V_C^1, w \in V_R^1\}$. In other words, we require that the (ordered) set of all edges incident in G to a vertex from V_R^1 must appear in G^1. Therefore, a subgraph is completely specified by its vertex set.

In particular, if $A \subseteq V_C$:

- The *subgraph spanned by A in G*, denoted as $\lceil A \rceil^G$, has $V_C(\lceil A \rceil^G) = A \cup N_G(N_G(A))$ and $V_R(\lceil A \rceil^G) = N_G(A)$.
- The *subgraph generated by A in G*, denoted as $\lfloor A \rfloor_G$, has $V_C(\lfloor A \rfloor_G) = A$ and $V_R(\lfloor A \rfloor_G) = \{v \in N_G(A) | N_G(v) \subseteq A\}$.

– For $A \subseteq V_R$, the subgraph induced by A in G, denoted $[A]_G$, has $V_C([A]_G) = N_G(A)$ and $V_R([A]_G) = A$.

Example. Let us consider $F = \{man(Bob), woman(Alice), loves(Bob, Alice)\}$ a fact. The bipartite graph representation $G = (V_C, V_R; E_G)$ consists of $V_C = \{man(Bob), woman(Alice)\}, V_R = \{loves(Bob, Alice)\}$ and E_G the corresponding edges linking Bob to $Alice$ via $loves$.

3.1 Semantics

Definition 3 (Interpretation). *Let $W = (P, C)$ be a vocabulary. An interpretation of W is a pair $I = (\Delta, .^I)$ where Δ is the domain of the interpretation, and $.^I$ a function where:* $\forall\ c$ *in* $C, c^I \in \Delta$ *and* $\forall\ p$ *in* $P, p^I \subseteq \Delta^{arity(p)}$.

An interpretation is non empty and can be possibly infinite.

Definition 4 (Model). *Let F be a fact on W, and $I = (\Delta, .^I)$ be an interpretation of W. I is a model of F iff there exists an application v:*

– $\forall\ c \in csts(F),\ v(c) = c^I$ *and*
– $\forall\ p(t_1,...,t_k) \in atoms(F),\ (v(t_1),...,v(t_k)) \in p^I$.

Definition 5 (Fact to logical formula). *Let F be a fact. $\phi(F)$ is the logical formula that corresponds to the conjunction of atoms in F. And $\Phi(F)$ corresponds to the existential closure of $\phi(F)$.*

Example. Let us consider a fact $F = \{person(x), name(x, Bob), age(x, 25)\}$.

– $\phi(F) = person(x) \wedge name(x, Bob) \wedge age(x, 25)$.
– $\Phi(F) = \exists x\ person(x) \wedge name(x, Bob) \wedge age(x, 25)$.

Property 1 (Model equivalence). Let F be a fact and I be an interpretation of W. Then I is a model of F iff I is a model (in the FOL sense) of $\Phi(F)$.

Definition 6 (Entailment). *Let F and G be two facts, F entails G if every model of F is also a model of G. The entailment relation is then noted $F \models G$.*

Definition 7 (Homomorphism). *Let F and F' be facts. Let $\sigma\colon terms(F) \to terms(F')$ be a substitution, i.e. a mapping that preserves constants (if $c \in C$, then $\sigma(c) = c$). We then note $\sigma(F)$ the fact obtained from F by substituting each term t of F by $\sigma(t)$. Then σ is a homomorphism from F to F' iff the set of atoms in $\sigma(F) \subseteq F'$.*

Example. Let $F = \{man(x_1)\}$ and $F' = \{man(Bob), woman(Alice)\}$. Let $\sigma : terms(F) \to terms(F')$ be a substitution such that $\sigma(x_1) = Bob$. Then σ is a homomorphism from F to F' since the atoms in $\sigma(F)$ are $\{man(Bob)\}$ and the atoms in F' are $\{man(bob), woman(Alice)\}$.

Property 2 (Entailment). Let F and Q be facts. $F \models Q$ iff there exists Π an homomorphism from Q to F.

In [18] homomorphism is denoted as projection and it is the fundamental operation on simple conceptual graphs. If we consider the bipartite depiction of facts mentioned above, a *projection from G to H* is a mapping $\Pi : V_C(G) \cup V_R(G) \to V_C(H) \cup V_R(H)$ such that:

- $\Pi(V_C(G)) \subseteq V_C(H)$ and $\Pi(V_R(G)) \subseteq V_R(H)$;
- $\forall c \in V_C(G),\ \forall r \in V_R(G)$ if $c = N_G^i(r)$ then $\Pi(c) = N_H^i(\Pi(r))$;
- $\forall v \in V_C(G) \cup V_R(G)\ \lambda_G(v) \geq \lambda_H(\Pi(v))$ where λ is a labelling of nodes with elements from a set of finite partially ordered set (the terminology, the support - please see next section.).

A projection $G \to H$ exists if and only if G *entails* H (i.e. $G \geq H$). Subsumption checking is an NP-complete problem [9].

3.2 Rules

Rules are objects used to express that some new information can be inferred from another information. Rules are also used in order to define the terminological knowledge corresponding to a set of facts. In the rules we could include the hierarchy of concept types, the hierarchy of relations and more complicated rules that do not simply define generalisations specialisations of types.

Rules are built from two different parts called head and body. Once the body of a rule can be deduced from a fact, then the information in the head should also be considered when accessing information. Please note that the head could contain new variables not present in the body. In this case applying such rules should be done with care as it can generate an infinite number of new facts to be taken into account.

Definition 8 (Rule). *Let H and B be facts. A rule is a pair $R = (H, B)$ of facts where H is called the head of the rule and B is called the body of the rule. A rule is commonly noted $B \to H$.*

Definition 9 (Rule model). *Let W be a vocabulary, I an interpretation on W, and R a rule on W. We say that I is a model of R iff for every justification V_B of B in I there exists a justification V_H of H in I such that $\forall t \in vars(B) \cap vars(H), V_B(t) = V_H(t)$.*

Definition 10 (Rule to logical formula). *Let $R = (H, B)$ be a rule. Let b_x be the variables from B, and h_x be the variables from H that are not in B, the logical formula corresponding to R is the following: $\Phi(R) = \forall b_x\ (\phi(B) \to \exists h_x\ \phi(H))$.*

Example. Let us consider a rule $R = \{person(x), person(y), sibling(x, y)\} \to \{person(z), parent(x, z), parent(y, z)\}$. $\Phi(R) = \forall x, y\ (person(x) \wedge person(y) \wedge sibling(x, y) \to \exists z\ person(z) \wedge parent(x, z) \wedge parent(y, z))$.

Property 3 (Model equivalence). Let R be a rule and I be an interpretation of W. Then I is a model of R iff I is a model (in the FOL sense) of $\Phi(R)$.

In conceptual graphs, the rules that define the terminology (hierarchy of concepts and relations) is defined in the so called support of the conceptual graph. The support is taken into account when performing projection (as explained in the previous section). In this paper we will also use the hierarchy of rules and concepts in the multi scale representation of knowledge. Therefore, we remind here the notion of support that will be used later on in the hierarchical knowledge base definitions.

The support is a tuple $S = (T_C, T_R)$ where T_C is a finite partially ordered set (poset), $(T_C, \leq)$, of *concept types*, defining a type hierarchy which has a greatest element $\top_C$, namely the universal type. In this specialisation hierarchy, $\forall x, y \in T_C, x \leq y$ is used to denote that x is a subtype of y. T_R is a finite set of *relation types* partitioned into k posets $(T_R^i, \leq)_{i=1,k}$ of relation types of arity i ($1 \leq i \leq k$), where k is the maximum arity of a relation type in T_R. Moreover, each relation type of arity $i, r \in T_R^i$, has an associated *signature* $\sigma(r) \in \underbrace{T_C \times \ldots \times T_C}_{i \text{ times}}$, which specifies the maximum concept type of each of its arguments. This means that if we use $r(x_1, \ldots, x_i)$, then x_j is a concept with $type(x_j) \leq \sigma(r)_j$ ($1 \leq j \leq i$). The partial orders on relation types of the same arity must be *signature-compatible*, i.e. it must be such that $\forall r_1, r_2 \in T_R^i \ r_1 \leq r_2 \Rightarrow \sigma(r_1) \leq \sigma(r_2)$. The sets T_C, T_R are mutually disjoint.

Before defining formally what a knowledge base is, let is make a note about how the support is related to facts. If we consider a fact $G = (V_C, V_R; E_G)$ then the nodes in the graph (fact), the concepts and the relation will respect signature wise the support (hierarchy given by the concept types and relation types). Formally, we can consider λ is a labelling of the nodes of G with elements from the support $S = (T_C, T_R)$: $\forall r \in V_R, \lambda(r) \in T_R^{d_G(r)}$; $\quad \forall c \in V_C, \lambda(c) \in T_C$ such that if $c = N_G^i(r), \lambda(r) = t_r$ and $\lambda(c) = t_c$, then $t_c \leq \sigma_i(r)$.

3.3   Knowledge Base

Definition 11 (Knowledge base). *Let W be a vocabulary. A knowledge base (KB) is a pair $K = (F, \mathcal{R})$ where F is a fact on W and $\mathcal{R}$ is a set of rules on W.*

Definition 12 (KB model). *Let $K = (F, \mathcal{R})$ be a knowledge base and I be an interpretation. I is a model of K iff I is a model of F and also a model of every rule R_i in $\mathcal{R}$.*

Definition 13 (Entailment). *Let K be a knowledge base and Q be a fact. K entails Q iff all models of K are also models of Q.*

Definition 14 (Logical representation). *Let $K = (F, \mathcal{R})$ be a knowledge base. $\Phi(K) = (\Phi(F), \Phi((R)))$ is the logical representation of K. $\Phi(F)$ is the logical formula of F and $\Phi(\mathcal{R}) = \bigcup_{r \in \mathcal{R}} \Phi(r)$.*

Property 4 (Model equivalence). Let K be a knowledge base and I be an interpretation. Then I is a model of K iff I is a model (in the FOL sense) of $\Phi(K)$.

In other words, given a knowledge base K and a conjunctive query Q, the RBDA problem consists in answering if Q can be deduced from K, denoted $K \models Q$.

Rule application can be performed of two different methods, called FORWARD CHAINING and BACKWARDS CHAINING.

Definition 15 (Applicable rule). *Let $R = (H, B)$ be a rule and F be a fact. R is applicable to F if there exists an homomorphism $\Pi : B \to F$. In this case, the application of R to F according to Π is a fact $\alpha(F, R, \Pi) = F \cup \Pi^{safe}(H)$.*

Please note the use of Π^{safe} instead of Π. Π^{safe} is an application that converts existential variables into fresh ones at the moment of joining new information with the initial fact. Such process is important in order to avoid unnecessary specializations. A derivation is the result of a finite sequence of rules application.

Definition 16 (Derivation). *Let F be a fact. F' is a derivation of F iff there exists a finite sequence of facts $F = F_0, ..., F_k = F'$ (called the derivation sequence) such that for every i there exists R and Π such that $F_i = \alpha(F_{i-1}, R, \Pi)$.*

Definition 17 (Saturation). *Let F be a fact and R be a set of rules. $\Pi_R(F) = \{\Pi : B_R \to F\}$ is the set of homomorphisms of the body of applicable rules to F. $\alpha(F, R) = F \bigcup_{\pi \in \Pi_R(F)} \pi^{safe}(H_R)$ is the result of the application of all those rules. The saturation of a fact is the process of applying rules from the initial fact until no more new information can be added to the fact via rule application. Let the initial fact $F_0 = F$, and $F_i = \alpha(F_{i-1}, R)$, a fact is saturated when $F_i \equiv F_{i+1}$.*

Theorem 1 (Equivalence). *Let $K = (F, R)$ be a knowledge base and Q be a fact. The following assertions are all equivalent:*

- $K \models Q$
- *there exists a derivation $F ... F'$ such that $F' \models Q$*
- *there exists an $n \in \mathbb{N}$ such that $F_n \models Q$*

When there are no rules in the ontology, the problem is then equivalent to homomorphism computation, which is a NP-hard problem. In the presence of rules, the problem is undecidable. Both the forward chaining and backwards chaining mechanisms are not certain of halting. This is easy to verify through the means of very simple examples.

Forward chaining. Let $K = (F, R)$ be a knowledge base, $F = person(Bob)$, and $R = \{\{person(x)\} \to \{parent(y, x), person(y)\}\}$.

Let $Q = \{parent(x, Tom)\}$ be a fact. Asking a forward chaining mechanism if Q can be deduced from K may eventually never stop. The mechanism will first verify if Q can be deduced from F, if there is an x having Tom as parent in F. As it is not the case, rules will be applied and F will be enriched into $F' = \{person(Bob), parent(p_1, Bob), person(p_1)\}$. The mechanism will then verify if Q can be deduced from F'. As it is still not the case, it will once again apply rules and enrich F' into F''. And it will do it infinitely as in this case, no answer will be ever found to the query.

Backwards chaining. Let $K = (F, R)$ be a knowledge base, and $R = \{\{p(x, y), p(y, z)\} \rightarrow \{p(x, z)\}\}$.

Let $Q = \{p(a, b)\}$ be a fact. Asking a backwards chaining mechanism if Q can be deduced from K may also eventually never stop. The mechanism will first verify if $\{p(a, b)\}$ can be deduced from F. If that is the case, the mechanism will stop. Otherwise, it will rewrite the initial query Q into a new query $Q_1 = \{p(a, x_0), p(x_0, b)\}$. Q is deduced from K if Q_1 is deduced from K. If Q_1 can not be deduced from F, the mechanism will rewrite the query again, for example with $Q_2 = \{p(a, x_0), p(x_0, x_1), p(x_1, b)\}$. Such sequence of rewritings may never end. Any finite rewriting corresponds to a finite sequence, for example, of length k of form $\{p(a, x_0) \ldots p(x_k, b)\}$. The facts could always contain a sequence of length $k + 1$.

4 Multi Scale Representation and Reasoning

In this section we introduce the notion of layered graphs that form the basis of our proposal for multi scale knowledge representation and reasoning. We detail the syntax and the semantics of layered graphs as well as a graph based projection inspired algorithm.

4.1 Representing Multi Scale Knowledge

We introduce the concept of a "complex node" that intuitively will be the node that will be expanded to generate the layers in the hierarchical representation. The complex nodes will only be concept nodes but their neighbours (relation nodes) will also be expanded. In order to formalise the transition from one level to the other, we introduce transitional descriptions.

Definition 1. *Let $G = (V_C, V_R; E_G)$ be a bipartite graph. A transitional description associated to G is a pair $\mathcal{TD} = (D, (G.d)_{d \in D \cup N_G(D)})$ where*

- *$D \subseteq V_C$ is a set of complex nodes.*
- *For each $d \in D \cup N_G(D)$ $G.d$ is a bipartite graph.*
- *If $d \in D$ then $G.d$ is the description of the complex node d. Distinct complex nodes $d, d' \in D$ have disjoint descriptions $G.d \cap G.d' = G^{\emptyset}$.*
- *If $d \in N_G(D)$ then either $G.d = G^{\emptyset}$ or $G.d \neq G^{\emptyset}$ and, in this case, $N_G(d) - D \subseteq V_C(G.d)$ and $V_C(G.d) \cap V_C(G.d') \neq \emptyset$ if and only if $d' \in N_G(d) \cap D$.*

A transitional description of a bipartite graph G provides a set D of complex nodes in one of the classes of the bipartition, each complex node having associated a description. This descriptions are disjoint bipartite graphs.

The neighbors of complex nodes either have empty descriptions or are described as bipartite graphs. These bipartite graphs contain in one of the classes

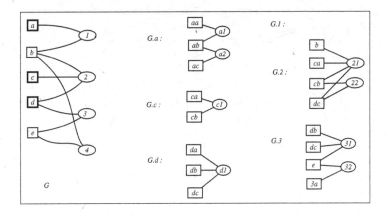

Fig. 1. Transitional description

of the bipartition, (V_C), all the atomic neighbors of the initial graph. The remaining nodes in each of these classes are new nodes or are taken from the descriptions of the corresponding complex neighbors of the initial graph.

Definition 2. *If $\mathcal{TD} = (D, (G.d)_{d \in D \cup N_G(D)})$ is a transitional description associated to the bipartite graph $G = (V_C, V_R; E_G)$, then the graph $\mathcal{TD}(G)$ obtained from G by applying $\mathcal{TD}$ is constructed as follows:*

1. *Take a new copy of $\lfloor V_C - D \rfloor_G$.*
2. *For each $d \in D$, take a new copy of the graph $G.d$ and make the disjoint union of it with the current graph constructed.*
3. *For each $d \in N_G(D)$, identify the nodes of $G.d$ which are already added to the current graph (i.e. the atomic nodes of G that are neighbours of d and the nodes of $G.d'$ which appear in $G.d$). For each complex neighbour d' of d in G, add the remaining nodes of $G.d$ as new nodes in the current graph and link all these nodes by edges as described in $G.d$ (in order to have an isomorphic copy of $G.d$ as a subgraph in the current graph).*

Figure 1 shows an example of transitional description, where graph G has $V_C = \{a, b, c, d, e\}$, $V_R = \{1, 2, 3, 4\}$ and the set of complex nodes from V_C (shown as bold rectangles) is $D = \{a, c, d\}$. The description of these nodes, namely $G.a$, $G.c$, and $G.d$, follows the same rule of node labelling (i.e. rectangle nodes are denoted by letters and oval nodes are denoted by numbers) and has a prefix association. $N_G(D)$ is the set $\{1, 2, 3\}$ whose description is shown in Fig. 1. Note that the description associated with node 1 is the empty graph. The nodes of $V_C(G.2)$ are (i) the atomic node b of G, (ii) the nodes ca and cb from $G.c$, and (iii) the node dc from $G.d$, since $N_G(2) = \{b, c, d\}$, b is an atomic node of G and c, d are complex nodes of G. In the description of $G.3$, a new node $3a$ appears besides nodes e and db, dc.

Figure 2 illustrates this construction for graph G and for the transitional description of G depicted in Fig. 1.

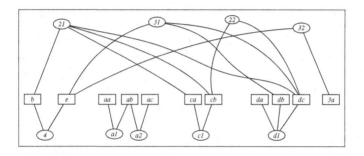

Fig. 2. The graph obtained from applying a transitional description

Theorem 2. *If $G = (V_C, V_R; E_G)$ is a bipartite graph and $\mathcal{TD}$ is a transitional description associated to G, then the graph $\mathcal{TD}(G)$ obtained from G by applying $\mathcal{TD}$ is also a bipartite graph.*

Proof. Let $H = \mathcal{TD}(G)$. By the construction of the graph H described in the previous definition, we have $V(H) = V_C(H) \cup V_R(H)$, where

$$V_C(H) = (V_C(G) - D) \cup \cup_{d \in D \cup N_G(D)} V_C(G.d)$$

and

$$V_R(H) = V_R(\lfloor V_C(G) - D \rfloor_G) \cup \cup_{d \in D \cup N_G(D)} V_R(G.d).$$

Furthermore, each edge of the graph H is either from $\lfloor V_C(G) - D \rfloor_G$ or from some bipartite description, hence has an endpoint in $V_C(H)$ and the other in $V_R(H)$.

Note that if $G.d = G^{\emptyset}$ for $d \in N_G(D)$, we have no description available for relation vertex d. This either depends on a lack of information or on an inappropriate expounding. The idea traces back to the notion of context in [18] or to the more elaborate notion of nested conceptual graph [10]. However, as our approach is not just a diagrammatic representation, the bipartite graph structure is taken into account.

This hierarchical representation allows, if we have a interconnected world described by a set of facts (a bipartite graph) and if we can provide details about both some complex concepts and their relationships, to construct a second level of knowledge about this world, describing these new details. This process can be similarly performed with the last constructed level, thus obtaining a coherent set of layered representations of the initial world. Please note that at each level, we need to highlight the specific set of hierarchical rules (concepts and rules) for that level. This means that at different levels we can use different terminologies that are specific to that level.

Definition 3. *Let d a nonegative integer. A hierarchical KB of depth d is a family* $\mathbf{HG} = \langle\, G^0, \mathcal{TD}^0, \ldots, \mathcal{TD}^{d-1}\,\rangle$ *where:*

- $G^0 = [S^0, (V_C^0, V_R^0; E^0)]$ *is a knowledge base composed of a support S and a set of facts $(V_C^0, V_R^0; E^0)$;*

- $\mathcal{TD}^0$ is a transitional description associated to G^0,
- for each $k, 1 \leq k \leq d-1, \mathcal{TD}^k$ is a transitional description associated to $G^k = [S^k, (V_C^k, V_R^k; E^k)] = \mathcal{TD}^{k-1}(G^{k-1})$.

We can define substructures of the HG model which can be useful to devise a customisable and versatile functionality that deals with large graph structures.

Definition 4. *If* $\mathbf{HG_1} = \{HG_1^0, \ldots, HG_1^{d_1}\}$ *and* $\mathbf{HG_2} = \{HG_2^0, \ldots, HG_2^{d_2}\}$ *are two hierarchical KBs, then* $\mathbf{HG_2}$ *is a subgraph of* $\mathbf{HG_1}$ *if* $d_2 \leq d_1$ *and* $\exists k, 0 \leq k \leq d_1 - d_2$, *such that for all* $i \in \{0, \ldots, d_2\}$: G_2^i *is a subgraph of* G_1^{k+i}, $D_2^i \subseteq D_1^{k+i}$ *and* $G_2.d$ *is a subgraph of* $G_1.d$ *for each* $d \in D_2^i \cup N_{G^i}(D_2^i)$.

4.2   Reasoning with Multi Scale Knowledge

Graph projection can be extended to hierarchical structures; however, with knowledge integration in mind, we only consider the case where queries are simple graphs (and not hierarchical graphs).

Definition 5. *A descending path of length* k *in* $HG = \langle G^0, \mathcal{TD}^0, \ldots, \mathcal{TD}^{d-1} \rangle$ *is a sequence* $P = v_0, \ldots, v_k$ $(k \leq d)$, *where* $v_i \in V(G^i)$ *and, for each* i, $1 \leq i \leq k$, *condition* $v_i \in V(G.v_{i-1})$ *holds. The last vertex of* P *is denoted as* $end(P)$. *Moreover* k, *i.e. the length of* P, *is denoted by* $length(P)$. *The set of all descending paths of* HG *is referred as* $\mathcal{P}(HG)$.

Definition 6. *Let* $HG = \langle G^0, \mathcal{TD}^0, \ldots, \mathcal{TD}^{d-1} \rangle$ *be a hierarchical knowledge base and* Q *a query. A hierarchical projection from* Q *to* HG *is a mapping* $\Pi : V_C(Q) \cup V_R(Q) \rightarrow \mathcal{P}(HG)$ *such that* $\forall v \in V(Q)$, *if* $\Pi(v) = P_v$:

- *if* $v \in V_C(Q)$, *then* $end(P_v) \in V_C(G^{length(P_v)}) - D^{length(P_v)}$ *and if* $v \in V_R(Q)$ *then* $end(P_v) \in V_R(G^{length(P_v)}) - N_{G^{length(P_v)}}(D^{length(P_v)})$;
- $\forall c \in V_C(Q), \forall r \in V_R(Q)$ *if* $c = N_G^i(r)$, *then* $length(P_c) \leq length(P_r)$ *and for each* v *on* P_r *at distance* k *from the start vertex of* P_r *such that* $length(P_c) \leq k \leq length(P_r)$, *we have* $N_{G^k}^i(v) = end(P_c)$.

If there is projection from Q to HG, then HG subsumes Q. Similar as explained in the previous section we an consider a logical semantics of the hierarchical knowledge bases and show its soundness with respect to graph operations. This semantics is sketched below.

Let $\mathbf{HG} = \langle G^0, \mathcal{TD}^0, \ldots, \mathcal{TD}^{d-1} \rangle$ be a hierarchical knowledge base and $S_{HG} = (T_C, T_R)$ be the union of the supports of its levels. A ternary predicate is assigned to each concept type from T_C, and an $n+1$-ary predicate is assigned to each relation type of arity n from T_R^n. Each predicate has the same name as the element of the support it is associated to. If $t \in T_C$, then the ternary predicate $t(x, y, z)$ holds. Intuitively, this means that (i) at level x, y is a concept vertex, (ii) the concept represented by this vertex is z, and (iii) its type is t. Similarly, if $t \in T_R^n$, then predicate $t(x, z_1, \ldots, z_n)$ holds. This means that (i) a relation vertex on the level x exists and that (ii) the relation represented by this vertex is $t(z_1, \ldots, z_n)$.

The formula $\Psi^*(HG)$ is constructed as the existential closure of x_k, where $0 \leq k \leq d - 1$ are the variables that represent the levels and $\Psi^*(G^k)$ be the formula obtained by adding x_k as the first argument of each member predicate for every level. Then,

$$\Psi^*(HG) = \exists x_0 \exists x_1 \ldots \exists x_{d-1} (\wedge_{k=0}^{d-1} \Psi^*(G^k))$$

Theorem 3. *Hierarchical projection is sound and complete with respect to Ψ^*.*

The proof is similarly to [11].

Since the semantics presented here are similar to the semantics of [11] a few words to compare the two formalisms are necessary. Transitional descriptions are a syntactical device which allows a successive construction of bipartite graphs. The knowledge detailed on a level of a hierarchy is put in context by using descriptions for relation nodes as well, while [11] only details the concept nodes and thus can be viewed as a particular instance of the formalism shown here.

5 Conclusion

This paper presented a transformation system that could be an appropriate hierarchical model for real world applications that require consistent transformation at different granularity levels. We presented the syntax of the extension and then demonstrated that the syntactic extension proposed is accompanied by sound and complete reasoning mechanisms.

Future work will focus on the application of such formalisation on a clear used case issued from life sciences. We are interested to see if the syntactic extension (i.e. the transitional descriptions) are easily elicited from non computing end users. To this end specific interfaces must be carefully devised in order to easily capture such knowledge.

The sound and complete reasoning mechanisms based on graphs are of great importance in a practical setting. We are thus interested to see how the projection operation can be best visualised for non computing experts. If demonstrated, this will be one of the main salient points of using such expressive formalism based on graph based syntax as opposed to approaches such as [6].

Acknowledgements. The second author acknowledges the support of ANR grants ASPIQ (ANR-12-BS02-0003), QUALINCA (ANR-12-0012) and DURDUR (ANR-13-ALID-0002). The work of the second author has been carried out part of the research delegation at INRA MISTEA Montpellier and INRA IATE CEPIA Axe 5 Montpellier.

References

1. Akyildiz, I.F., Su, W., Sankarasubramaniam, Y., Cayirci, E.: A survey on sensor networks. IEEE Commun. Mag. **40**(8), 102–114 (2002)
2. Baader, F.: Logic-based knowledge representation. In: Wooldridge, M.J., Veloso, M.M. (eds.) Artificial Intelligence Today. LNCS (LNAI), vol. 1600, pp. 13–41. Springer, Heidelberg (1999)

3. Baader, F., Brandt, S., Lutz, C.: Pushing the EL envelope. In: Proceedings of IJCAI 2005 (2005)
4. Baget, J.-F., Croitoru, M., da Silva, B.P.L.: ALASKA for ontology based data access. In: Cimiano, P., Fernández, M., Lopez, V., Schlobach, S., Völker, J. (eds.) ESWC 2013. LNCS, vol. 7955, pp. 157–161. Springer, Heidelberg (2013)
5. Baker, C.J., Cheung, K.-H.: Semantic Web: Revolutionizing Knowledge Discovery in the Life Sciences. Springer Science & Business Media, New York (2007)
6. Benczúr, A., Hajas, C., Kovács, G.: Datalog extension for nested relations. Comput. Math. Appl. **30**(12), 51–79 (1995)
7. Calì, A., Gottlob, G., Lukasiewicz, T.: A general datalog-based framework for tractable query answering over ontologies. Web Semant. Sci. Serv. Agents World Wide Web **14**, 57–83 (2012)
8. Calvanese, D., De Giacomo, G., Lembo, D., Lenzerini, M., Rosati, R.: Tractable reasoning and efficient query answering in description logics: the DL-Lite family. J. Autom. Reason. **39**(3), 385–429 (2007)
9. Chein, M., Mugnier, M.-L.: Conceptual graphs: fundamental notions. Revue d'Intelligence Artificielle **6**–**4**, 365–406 (1992)
10. Chein, M., Mugnier, M.-L.: Positive nested conceptual graphs. In: Lukose, D., Delugach, H.S., Keeler, M., Searle, L., Sowa, J.F. (eds.) ICCS 1997. LNCS, vol. 1257, pp. 95–109. Springer, Heidelberg (1997)
11. Chein, M., Mugnier, M.L., Simonet, G.: Nested graphs: a graph-based knowledge representation model with fol semantics. In: Anthony, S.C.S., Cohn, G., Schubert, L.K. (eds.) KR, pp. 524–535. Morgan Kaufmann, San Francisco (1998)
12. Croitoru, M., Compatangelo, E., Mellish, C.: Hierarchical knowledge integration using layered conceptual graphs. In: Dau, F., Mugnier, M.-L., Stumme, G. (eds.) ICCS 2005. LNCS (LNAI), vol. 3596, pp. 267–280. Springer, Heidelberg (2005)
13. Croitoru, M., Xiao, L., Dupplaw, D., Lewis, P.: Expressive security policy rules using layered conceptual graphs. Knowl. Based Syst. **21**(3), 209–216 (2008)
14. Da Silva, B.P.L., Baget, J.-F., Croitoru, M.: A generic platform for ontological query answering. In: Bramer, M., Petridis, M. (eds.) Research and Development in Intelligent Systems XXIX, pp. 151–164. Springer, London (2012)
15. Esch, J., Levinson, R.: An implementation model for contexts and negation in conceptual graphs. In: Ellis, G., Rich, W., Levinson, R., Sowa, J.F. (eds.) ICCS 1995. LNCS, vol. 954, pp. 247–262. Springer, Heidelberg (1995)
16. Handfield, R.B., Nichols, E.L.: Introduction to Supply Chain Management, vol. 1. Prentice Hall Upper, Saddle River (1999)
17. Mugnier, M.-L.: Ontological query answering with existential rules. In: Rudolph, S., Gutierrez, C. (eds.) RR 2011. LNCS, vol. 6902, pp. 2–23. Springer, Heidelberg (2011)
18. Sowa, J.: Conceptual Structures: Information Processing in Mind and Machine. Addison-Wesley, Reading (1984)
19. Wache, H., Voegele, T., Visser, U., Stuckenschmidt, H., Schuster, G., Neumann, H., Hübner, S.: Ontology-based integration of information-a survey of existing approaches. In: IJCAI 2001 Workshop: Ontologies and Information Sharing, vol. 2001, pp. 108–117. Citeseer (2001)

Transforming UML Models to and from Conceptual Graphs to Identify Missing Requirements

Bingyang Wei[1(✉)] and Harry S. Delugach[2]

[1] Computer Science Department, Midwestern State University,
Wichita Falls, TX 76308, USA
bingyang.wei@mwsu.edu
[2] Computer Science Department, University of Alabama in Huntsville,
Huntsville, AL 35899, USA
delugach@cs.uah.edu

Abstract. For a set of UML models built for a system, ensuring that each model is relatively complete with respect to the rest of the set is critical to further analysis and design. In this paper, we present a novel idea to identify requirements gap in a model by synthesizing requirements from other types of models. This is accomplished by a bidirectional transformation between a set of partially complete UML models and conceptual graphs. UML models are first transformed to conceptual graphs based on a well-defined set of primitives and canonical graphs to form a centralized requirements knowledge reservoir, then inference rules are applied to unveil possible missing requirements. These identified missing requirements, when transformed back to UML, can provide prompts of requirements gap to software modelers and stimulate them to come up with more requirements in order to resolve them, thus making UML models more complete.

Keywords: Knowledge representation and reasoning · Requirements acquisition · UML · Conceptual graphs

1 Introduction

During requirements process, various types of UML models are built, each of which captures a specific view and all models together constitute the overall description of the system under development. This multiple-viewed requirements modeling approach relies on a set of interdependent models. However, it is difficult for a modeler to know whether a model is complete or what requirements are missing in the current version of the model. An effective way of finding more requirements for a model is by looking at other types of models that are being developed in parallel for the system under development: missing requirements of a model can be identified by synthesizing requirements from other types of models.

© Springer International Publishing Switzerland 2016
O. Haemmerlé et al. (Eds.): ICCS 2016, LNAI 9717, pp. 72–79, 2016.
DOI: 10.1007/978-3-319-40985-6_6

Based on that idea, we propose a bidirectional transformation approach to identify missing requirements in a UML model. This approach adopts conceptual graphs (CGs) [6]. Since the models of a software system can be regarded as a collection of statements that evaluate to truth, i.e. assertional knowledge, CGs are appropriate for representing software requirements. Other research papers that have adopted CGs to represent requirements are [4,5]. Figure 1 illustrates the bidirectional transformation approach for the purpose of identifying missing requirements for UML models.

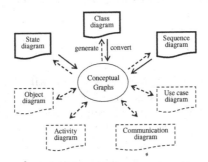

Fig. 1. Bidirectional transformation for missing requirements identification.

UML models are first converted to CGs to form a centralized requirements knowledge reservoir, which is called CGs Reservoir. If a modeler needs feedback about completeness of a particular model, all requirements in the CGs Reservoir are transformed back to UML notation of that particular model. Note that requirements from other types of UML models also get transformed and represented in that particular model. The modeler will be presented with a model with new requirements that she is not aware of before transformation. Moreover, these newly generated requirements can be used as stimulations and encourage the modeler to come up with more requirements about the model. For example, the way to identify missing requirements in a partially complete class diagram is to transform requirements in other models to class diagram through CGs. Seeing new requirements being added in the original class diagram, the modeler will start adjusting and supplementing the new model. After this supplementing process, this augmented class diagram will in turn affect other models, causing similar effects in other models. Due to space limitations of this paper, only class diagrams, state diagrams, and sequence diagrams are considered here (solid shapes in Fig. 1). The structure, state, and interaction views provide a sufficiently broad range of the semantics of object-oriented models to show the generality of the approach.

The rest of the paper is structured as follows. Section 2 provides detailed explanations of a key component called CGs Support, based on which the bidirectional transformation is made possible. Transforming UML diagrams to and from CGs for the purpose of identifying missing requirements is fully described in Sect. 3 using a simple example. In Sect. 4, we conclude the paper and point to the future work.

2 CGs Support

The CGs Support is a key component in our work. It defines, in CGs form, semantic elements called primitives as building blocks of the three types of UML diagrams (class, state, and sequence diagrams), canonical graphs, which are used to express the three UML diagrams in CGs, and inference rules for identifying missing requirements.

2.1 Primitives

Software engineering researchers have tried to identify the minimal set of fundamental elements that underlies the requirements of an object-oriented system [2,3]. In the light of their work, the CGs Support defines a set of elemental concepts and relations underlying the three UML diagrams so that any requirement captured by the three UML diagrams can be expressed in terms of the primitives. The primitives are to UML modeling languages as assembly language statements are to high-level programming languages. The types of primitive concepts and relations are organized in a CGs concept type hierarchy and a CGs relation type hierarchy (Fig. 2(a) and (b)), respectively.

The *Object* type is the general description of something that has states and behaviors. The *Activity* type is used to describe all behaviors performed by objects. For example, enrolling in a seminar and taking exams in a university information system are both instances of the type *Activity*. The *Action* type describes the smallest computation unit, such as issuing and receiving messages. The *Message* type describes information exchanged between objects. It has two subtypes, *CallMessage* and *SignalMessage*. The *Time* and *Signal* describe time and signals, respectively.

Primitive relations relate primitive concepts in order to represent meaningful relationships among them. The types of primitive relations in the CGs Support are organized in a CGs relation type hierarchy. Figure 2(c) to (h) show some meaningful CGs snippets composed of primitive concepts and relations.

An *attribute* relation relates an *Object* type concept to a *T* type concept (Fig. 2(c)); the referent of the *T* type concept denotes an attribute value of the object represented by the *Object* type concept. An example of this is *Color: red* or *Size: large*. An *association* relation relates two *Object* type concepts (Fig. 2(d)), and it represents the semantic relationship that can occur between two objects. For example, a Student object is associated with a Seminar object by enrolledIn association. An *operation* relation relates an *Object* concept to an *Activity* type concept, and it depicts the relationship between an object and its operation. The CGs in Fig. 2(e) means that an object has the ability of performing a certain operation. An *agent* relation relates an *Object* type concept to an *Activity* type concept (Fig. 2(f)), which also describes the relationship between an object

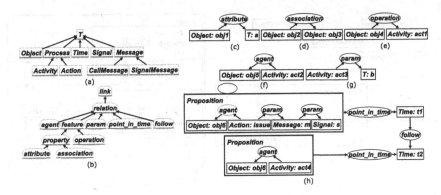

Fig. 2. Primitive concept type hierarchy, relation type hierarchy, and meaningful CGs made up of primitive concepts and relations.

and an operation, but it means that the operation is actually performed by an object. A *param* relation is used to associate a parameter to the main part (Fig. 2(g)), such as the arguments of an operation and the content of a message. A *point_in_time* relation relates a *Proposition* context to a *Time* type concept. A *follow* relation connecting two situations expresses the meaning of one starts after the previous situation is done (Fig. 2(h)).

2.2 Canonical Graphs

Each type of UML diagrams has a set of basic model elements. For example, class diagrams include a set of basic model elements such as classes, attributes, operations, and associations. In this work, for each type of UML diagrams, a set of canonical graphs is developed in which the basic model elements are properly represented in terms of the primitive concepts and relations as introduced in Sect. 2.1. By instantiating the set of canonical graphs, the requirements knowledge of a UML diagram is represented in CGs. In this way, a UML diagram is converted to CGs. In this section, canonical graphs of class diagrams, state diagrams, and sequence diagrams are presented.

The canonical graph for a class is shown in Fig. 3(a). Since a class describes a set of similar objects, the concept *ClassName: @forall* means "for all objects of this class." In this concept, *ClassName* denotes the name of a class and will be replaced with the name of a real class when this canonical graph is instantiated. An attribute is represented by a *T* type concept, which is related to the class concept through an *attribute* relation, while an operation is represented by an *Activity* type concept, which is related to the class concept by an *operation* relation. The *association* relates each object of this class to other objects. In Fig. 3(a), only one attribute, one operation, and one association are shown in the canonical graph. The canonical graph in Fig. 3(b) represents generalization.

A state diagram can be considered as a set of state transitions. The canonical graph in Fig. 3(d) represents "An object is in a state performing an activity while

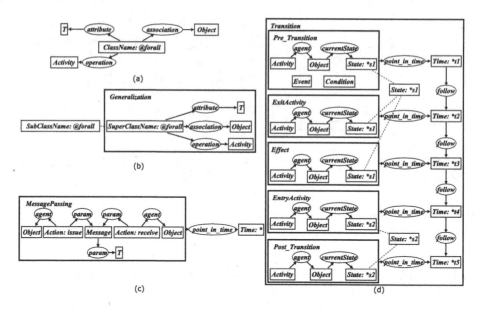

Fig. 3. Canonical graphs for class, state and sequence diagrams.

an event occurs at time t1; the guard condition is also satisfied, so the object performs exit activity at time t2 and performs effects on the transition at time t3; before it enters the second state at time t5, the object performs the entry activity of the second state at time t4." In the canonical graph of a state transition, object owning the state machine, effects on transitions, entry/exit, and do activities of states are already represented by primitive concepts such as *Object* and *Activity*. However, basic model elements like events, states, and guards are still not expressed in primitive forms (see *Event*, *State*, and *Condition* concepts in Fig. 3(d)). More canonical graphs for those model elements are available in [7].

Based on the semantics of a sequence diagram, its CGs are composed of a sequence of *MessagePassing* contexts. The canonical graph in Fig. 3(c) describes the semantics of a message exchange between objects at a certain point in time. A *Message* can carry parameters.

2.3 Inference Rules

Besides primitives and canonical graphs, the CGs Support also contains rules, which are used to infer requirements knowledge. Our approach is based on a forward-chaining inference method. As in any logical inference, the presence of a rule's antecedent in the CGs Reservoir implies its consequent, which represents the desired requirements knowledge used to build a target UML diagram. In this work, for each type of UML diagrams, a set of CGs inference rules is defined in order to derive requirements knowledge from the CGs Reservoir to build UML

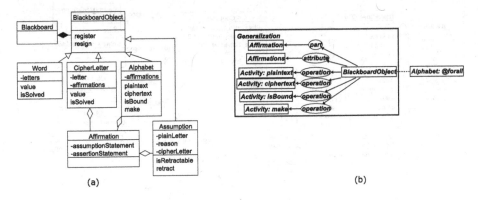

Fig. 4. Cryptanalysis class diagram and one class in CGs.

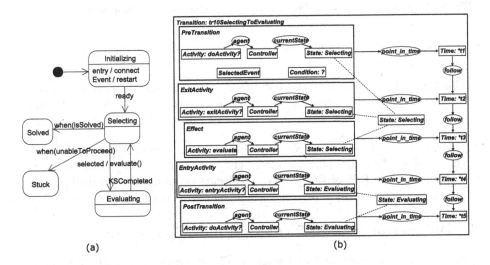

Fig. 5. Controller's state diagram and one state transition in CGs.

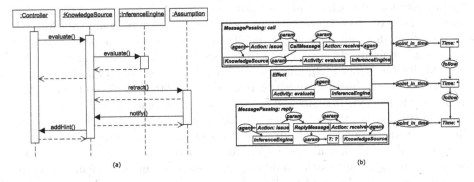

Fig. 6. Cryptanalysis sequence diagram and some messages in CGs.

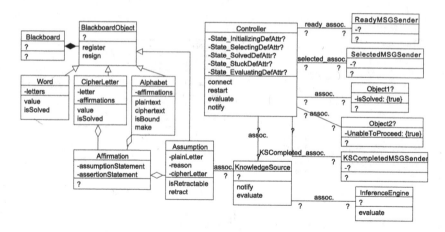

Fig. 7. Generated UML class diagram with requirements acquisition opportunities.

diagrams. Due to the size of CGs, we only present some of the inference rules for class diagrams in English:

- Class rule: If an *Object* type concept appears, then its class needs to be defined;
- Attribute rule: If an *Object* type concept is related to a concept *y through *attribute*, then the concept *y is an attribute value of the class of the *Object* type concept;
- Operation rule: If an *Object* type concept performs a *Process* type concept *y, then concept *y is an operation of the class of the *Object* type concept;
- Association rule 1: If two *Object* type concepts communicate through passing messages, then they have an *association* relation between them.

3   Bidirectional Transformation in a Case Study

This case study includes the UML diagrams of a system dealing with deciphering cryptograms [1]. The three partially complete UML diagrams are first converted to CGs (Figs. 4, 5 and 6) according to their canonical graphs. Because of space limitations, not all elements in the diagrams are converted. After applying the inference rules of class diagrams, the CGs in the CGs Reservoir are transformed back to class diagram notations (Fig. 7), the class diagram modeler will find those new requirements (gray classes and question marks) that were not in the original diagram (Fig. 4(a)). This generated UML classes diagram clearly provides several requirements acquisition opportunities: attributes need to be added in the previously existing and new classes, and associations need to be specified between previously existing classes and new ones. After the modeler resolves all issues in the generated class diagram, this more complete class diagram will be converted to CGs again, so the CGs Reservoir becomes more complete than before. State diagram modelers and sequence diagrams modelers can pull their diagrams out of the central CGs Reservoir in a similar manner. That is to say, another requirements discovering and completing iteration starts.

4   Conclusion and Future Work

In this work, by transforming UML diagrams to and from CGs, requirements acquisition opportunities are exposed. This approach is particularly useful for the requirements acquisition process among a team of requirements modelers preparing a software specification from different viewpoints. The current primitive concepts and relations work well for the three types of UML models. We do not claim the completeness of this set of primitives, since we have not considered all UML diagrams and other complex model elements. Another current limitation is the lack of automation support. Future work obviously will focus on automation and more formal evaluation.

References

1. Booch, G., Maksimchuk, R., Engle, M., Young, B., Conallen, J., Houston, K.: Object-Oriented Analysis and Design with Applications, 3rd edn. Addison-Wesley Professional, Reading (2007)
2. Dardenne, A., Van Lamsweerde, A., Fickas, S.: Goal-directed requirements acquisition. Sci. Comput. Program. **20**(1), 3–50 (1993)
3. Davis, A.M., Jordan, K., Nakajima, T.: Elements underlying the specification of requirements. Ann. Softw. Eng. **3**(1), 63–100 (1997)
4. Delugach, H.S.: An approach to conceptual feedback in multiple viewed software requirements modeling. In: Joint Proceedings of the Second International Software Architecture Workshop (ISAW-2) and International Workshop on Multiple Perspectives in Software Development (Viewpoints 1996) on SIGSOFT 1996 Workshops, pp. 242–246. ACM (1996)
5. Jaramillo, C.M.Z., Gelbukh, A., Isaza, F.A.: Pre-conceptual schema: a conceptual-graph-like knowledge representation for requirements elicitation. In: Gelbukh, A., Reyes-Garcia, C.A. (eds.) MICAI 2006. LNCS (LNAI), vol. 4293, pp. 27–37. Springer, Heidelberg (2006)
6. Sowa, J.F.: Conceptual Structures: Information Processing in Mind and Machine. Addison-Wesley, Menlo Park (1983)
7. Wei, B.: A comparison of two frameworks for multiple-viewed software requirements acquisition. Ph.D. thesis, University of Alabama in Huntsville (2015)

SDDNet: A Semantic Network for Software Design and Development Domain Via Graph Database

Shipra Sharma$^{(\boxtimes)}$ and Balwinder Sodhi

Indian Institute of Technology Ropar, Rupnagar 140001, India
{shipra.sharma,sodhi}@iitrpr.ac.in
http://www.iitrpr.ac.in

Abstract. When building software systems, deciding implementation details *manually* for specific system requirements is time consuming and often leads to sub-optimal choices. An automated, knowledge-base driven approach to select from possible technology alternatives can replace or reduce manual effort in realizing these requirements. This motivated us to develop such a knowledge system which can ultimately help in opportune and cost-effective software development. We create a rich knowledge-base by automatic extraction of *useful information* from a large volume of existing documentation pertaining to software components and technologies that are used to build variety of business applications. We store the knowledge-base in a graph database. A semantic network depicting relations between concepts found in Software Design and Development (SDD) domain is constructed from the database. This knowledge-base can be queried for deducing additional facts about the concepts stored therein.

Keywords: Knowledge system · Information extraction · Semantic network · Semantic computing · Software architecture · Software design

1 Introduction

A knowledge system is an organized composition of concepts with relevant and logical relations in a particular domain. A knowledge-intensive process like Software Design and Development (SDD) generates a colossal amount of information which is diverse and ever increasing. This information is most of the time unstructured, and less useful in a different SDD context. In such a scenario, a knowledge system, based on semantic network (in SDD domain), would help in converting unstructured information about an existing software component into various concepts. A semantic network visually represents the communication between interesting textual entities [4]. This system can be used to make rational decisions during SDD of other future software projects.

Software architecture and design is basically the "how" part of Software Development Life Cycle (SDLC) where decisions on packages, programming languages, program layers and modules and other such software engineering details

© Springer International Publishing Switzerland 2016
O. Haemmerlé et al. (Eds.): ICCS 2016, LNAI 9717, pp. 80–88, 2016.
DOI: 10.1007/978-3-319-40985-6_7

are made. A software architect has two options before developing an architectural solution of a problem: (i) start from scratch or (ii) learn from existing projects developed in same domain [12]. The second choice, however, is constrained by the lack of proper architectural documentation in the existing projects [3]. Also, if such documents do exist it takes a lot of manual effort to identify the exact required information. Our motivation is to assist in this decision process by way of providing queryable knowledge-base of SDD knowledge. Such a knowledge-base can be created by suitably extracting relevant information from textual documents accompanying software components used for building software application. A semantic network is proposed for SDD to depict the basic concepts and their realizations. In other words, we transform unstructured project descriptions into a direct queryable form, which can become a knowledge-base of any intelligent system in SDD domain.

The paper. is presented as follows. In Sect. 2, we discuss detailed development procedure of SDDNet and give a comprehensive discussion. Section 3 discusses evaluation results and their analysis of queries against our knowledge-base. Section 4 presents related work specific to knowledge representation in software architecture, design domain and development domain. To the best of our knowledge, there exist no knowledge system or conceptual structure for an all-inclusive SDD domain as SDDNet. Finally, Sect. 5 concludes our paper with an explanation of work in progress.

2   SDDNet: Design and Development Steps

Following steps give a detailed explanation of the methodology used to obtain SDDNet.

2.1   Controlled Vocabulary

We make use of a controlled vocabulary to construct our knowledge-base. This vocabulary has been developed for Software Engineering terms using standard text. The standard texts we consider are, [8,12] to generate this vocabulary. We used terms from the indexes of the terms given at end of each text. These indices are scraped to extract only words and leave the page numbers and any other irrelevant text. All the repeating terms are discarded. We believe that these two texts used by us give reasonable coverage of terms in software engineering domain and chances of missing relevant term are within acceptable limit. Our controlled vocabulary is currently populated with 2702 terms.

2.2   Relation Filter

We develop a set of relations (verb phrases) that are relevant in SDD domain. To find relations that are commonly used in SDD domain we use Stackoverflow[1]®. We specifically search Stackoverflow for architecture and design

---

[1] Well-known discussion forum.

questions w.r.t to software development phase. We scraped through top 3 rated answers of all such questions and extracted verbs from them. All such verbs which are used between architectural (and design) entity and its realization are stored in the set - Relation Filter (Ψ). A subset of Ψ is:

implements, written_in, has_properties, build_on, has, provides, instead_of, can_replace, associated, runs_on, requires, uses, facilitates, improves, optimizes, enhances, such_as, fixes, integrates, expedite, is_a, consists, based_on

2.3 The Complete Development Steps

Steps described are fully automated and can be achieved without any human intervention.

1. **Crawling and Pre-processing** - Documents related to software components available on various vendors (like IBM, Apache, etc.) are crawled. All the data in user manual and feature description of a software component are stored in one text document that uniquely relates to that component. Documents in the corpus are processed to remove all non-textual data (such as images). Further, sentences are corrected to have proper beginning and end as many of these documents do not have properly structured sentences. Following is the partial depiction of processed text document of our corpus. The software component it belongs to is Apache Hadoop.

The Apache Hadoop software library is a framework. It allows for the distributed processing of large data sets across clusters of computers using simple programming models. It is designed to scale up from single servers to thousands of machines, each offering local computation and storage. Rather than rely on hardware to deliver high availability, the library itself is designed to detect and handle failures at the application layer, so delivering a highly available service on top of a cluster of computers, each of which may be prone to failures.

2. **Information Extraction from Documents** - Each document in our plain corpus passes through the following two steps to give us a set of "prelim entity relations (Φ)".
   (a) **Pruning:** A sentence is pruned/discarded if none of their constituent words match a term from our controlled vocabulary (Sect. 2.1). This step may lead to some False Positives (FPs) and False Negatives (FNs)[2]. For verification, we randomly choose 500 sentences from our corpus and manually check the output of this step. For such selection FP rate was found to be 0.83 while FN rate was 0.17.
   (b) **Generating Entity-Relation Set (Φ):** We construct our unigram tagger (specific to SDD domain) with NLTK (Part-of-Speech) POS tagger [2] to attach POS with every word. Next, we define few grammatical rules to form separate chunks of nouns and their adjectives (and their forms) and verbs in a sentence. These rules are applied to chunk every sentence. Thus, we extract $\{Entity\#1, Relation, Entity\#2\}$ triples from each sentence forming a set Φ. We then store the extracted triples $\{Entity\#1, Relation, Entity\#2\}$ as one tuple in Φ. For example, processing of first sentence in box (depicted in *Step1* above) will

---

[2] FP- Those sentences which were falsely classified as belonging to SDD domain. FN- Those sentences which were falsely classified as not belonging to SDD domain.

add tuple - $\{Entity\#1 = Apache\ Hadoop\ Software\ Library, Relation = is\ a, Entity\#2 = framework\}$ to Φ (depicted in Eq. 1).

$$\Phi = \{.., \{..\}, \{Apache\ Hadoop\ Software\ Library, is\ a,$$
$$framework\}, \{..\}, \{..\}, ..\} \quad (1)$$

3. **From Prelim Entity Relations (Φ) to Terminal Relations (Φ')** - For each tuple $\phi \in \Phi$,: let ϕ_2 be the 2nd element of ϕ. If $\phi_2 \approx^3 \psi$, where $\psi \in \Psi$ we add ϕ to Terminal Relations (Φ'). Continuing with our running example, Φ in Eq. 1 has - "is a" as a 2nd element of one of it's tuple. Also, $is\ a \in \Psi$. Hence this tuple is added to Φ'. So,

$$\{Apache\ Hadoop\ Software\ Library, is\ a, framework\} \in \Phi' \quad (2)$$

4. **Processing Terminal Relations (Φ')** - Let ϕ'_1 and ϕ'_3 receptively be 1st and 3rd element of $\phi' \in \Phi'$. Then:
   (a) If ϕ'_1 or ϕ'_3 is not a single word we tag all its words with its part of speech using POS tagger [2]. If they all come out to be nouns we consider this element as "compound entity", else we separate the part which is noun and consider it as "single entity". All other parts become properties of this "single entity". Considering our running example 1st element of a tuple in Φ' is $Apache\ Hadoop\ Software\ Library$ (from Eq. 2). The POS tagger gives us [(Apache, NP), (Hadoop, NP), (Software, NN), (Library, NN)] and we conclude that this element is a compound entity.
   (b) We find it's (ϕ'_1 or ϕ'_3) *concept* by scraping Wikipedia's® search engine to find it's wiki page. From Eq. 2, for ϕ'_3 - "framework", our system populates its concept as - ['abstraction', 'generic functionality', 'reusable software']. Hence, framework is an instance of these concepts.
5. **Φ' to GraphDatabase** - We populate our graph database with the processed tuples of Φ'. The layout of our final graph database is shown in Fig. 1a. *Entity* nodes are 1st and 3rd element of each tuple of Φ' whereas, 2nd element of each tuple of Φ' are the self loop on *Entity* nodes. The *Software_Concept* labeled nodes are populated from data obtained in Step 4(b). The default value of *Software_Concept.name* is $Undefined\_Concept^4$. *Software_Component* labeled nodes represents all the software components which were crawled from the on-line portals. The property of these nodes are depicted below them. Figure 1b depicts a part of graph database for the running example of Eq. 2 after it has undergone all the processing steps.
6. **Network (SDDNet) via Graph Database** - Further on, we smooth our graph database to eliminate semantic gaps (if any) between nodes. An entity (e_1) is conceptually similar to another entity (e_2) if $e_1.name = e_2.name$. A relation - : is_same_as is formed between all such conceptually similar but not

---

[3] We make use of FuzzyWuzzy [10] and Wordnet [7] to avoid loss of relations similar (but not exactly same) to some relation in Ψ.

[4] All those entities for which we could not obtain any concept from Wikipedia we assign them a default concept - $Undefined\_Concept$.

related nodes. For example, we have a node with its *name.property* = *Hadoop* and it's label is *Software_Component*. Also, we have another node with its *name.property* = *Hadoop*, but it's label is *Entity*. Although these two nodes are conceptually same but they are not related in our graph database. Figure 2 partially depicts how this step creates relation for our (running) example (Fig. 1b). Following this logic for all entities, we complete the conceptual links of our graph database and form a complete semantic network and name it Software Design and Development Network (SDDNet).

7. **User Queries on SDDNet -** We use Neo4j [13] to store our knowledge-base as a graph. User can run queries against our knowledge-base in any of the Neo4j [13] supported query languages. For example, if a user desires to find all software components which are related by any distance to the *Software IDE* concept. In other words, software components which are somehow semantically related to IDE concept of software engineering are to be extracted. The output lists all *Software_Component* labeled nodes which link to software IDE through their entities. One row from output is: a concept class - {*IDE, software development environment*}, one of its instance is - {*Eclipse Java Editor*}, which is an entity of -{*IBM-Rational Functional Tester*} component. From this particular relation we can derive - *"Rational Functional Tester provided by IBM has a feature Eclipse Java Editor which in turn is an IDE."*

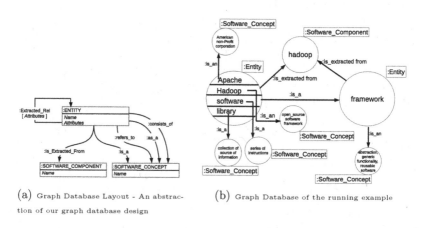

(a) Graph Database Layout - An abstraction of our graph database design

(b) Graph Database of the running example

Fig. 1. Graph database - (a) Abstraction (b) Actuality

3   Implementation and Analysis

This section presents the experimental environment and deductions and suggestions acquired based on SDDNet. In addition, we discuss threats to validity to the proposed layout.

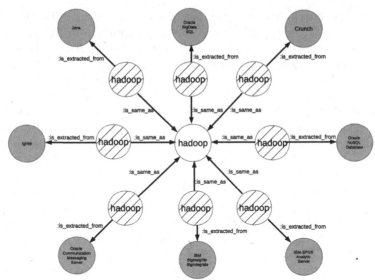

The center (unfilled) node is the particular node with *name=Hadoop* and *label=Software_Component*. All the pattern-filled nodes are nodes with *name* = *Hadoop* and *label=Entity*. Finally, all color-filled nodes are the software components to which the (pattern-filled) nodes belong as entity.

Fig. 2. Partial network for one software component (*Hadoop*)- After making additional semantic connections

3.1   Implementation Details

Input Data: We crawled 1001 software component descriptions from IBM®, 725 from Oracle® and all that are present on Apache®. **Knowledge-Base Data:** Using above data we obtained around $1,80,000$ tuples in the Prelim Entity Relations (Φ) set and around $40,300$ in Terminal Relation (Φ') set. The final semantic net consists of around $5,000$ nodes labeled under *Software_Concepts*, and total of 1715 *Software_Component*, exist whose properties populate our net. The current knowledge-base uses Neo4j [13] for storage.

3.2   Analysis and Discussion

SDDNet has been developed to assist in the process of choosing right software components to implement various architectural and design abstractions present in a software design. We analyze SDDNet's performance on (i) qualitative analysis and (ii) scenario-based output.

1. **Qualitative Analysis -** A qualitative analysis of semantic network is basically the quality of first order logic deductions that can be formulated from it [4]. Few deductions that have been derived from SDDNet are as follows. *Deduction 1-* A software component which uses another software component automatically inherits features and their attributes which are extracted from the latter component. For example (in Fig. 2), one of the software components (in outer-most layer) which inherits all the features (entities and their

attributes) of *Apache Hadoop* (the center node) is *Oracle Big Data Connectors*. Hence, it (*Oracle Big Data Connectors*) automatically becomes *highly available* as one of the features extracted from *Apache Hadoop* is *highly available*. *Deduction 2-* We can extract the realization-view[5] of our SDDNet. This is accomplished by reversing relations from software concepts to their realizations and its related properties. An example from SDDNet: Move from the concept *IDE* and reverse its relation : *is_a* to: *has*. This provides all the realization options of IDE present in SDDNet like Eclipse, JDeveloper and so on. Further navigation leads to their specific features like syntax checking, color coding etc. Similarly, other deductions can be reached and other views formed via SDDNet.

2. **Scenario-based Analysis -** Consider an architectural pattern and its solution as a scenario. Our objective is to estimate how relevant are the suggestions made by our system in terms of ground truths. We consider a 3-tier pattern and query SDDNet for each component of such a system. The queries are according to requirements and constraints of the scenario. Table 1 depicts the query results and the parameters on which they can further be modified. As can be observed in the Table 1, SDDNet gives relevant results for the pattern components.

3.2.1   Threats to Validity

Analysis provided by us have some threats to its validity. First, we do not consider all the projects or service providers. Hence the recommendations of SDDNet become biased towards the platforms chosen (in this study) and their products and services. Second, choice of grammatical rules defined by us to chunk entities influence the results. Third, the accuracy of analysis depends on accuracy of data used to built it. If a description of software component mentions that it is

Table 1. Architectural elements and their realization

Pattern component	Application required	SDDNet recommends	Filter properties
Client	IDE for client programming	Oracle JDeveloper, Apache Flex, IBM Rational Application Developer	Refactoring, Color Coding, Syntax Checking
Presentation layer	Web Server	IBM Web Sphere, Apache HTTP Server, Apache TomCat, Apache DeltaSpike	Uses ASP or JSP, OS support, Extensible
Business logic layer	Application-Server	IBM Web Sphere, Oracle JRockit Products, Apache DeltaSpike, IBM CICS Transaction Gateway, Oracle Web Logic Server	Reliability, Scalable, Security
Data layer	Database	Apache Giraph, Apache Lucene, Apache Accumulo, Oracle Database Mobile Server, Oracle Embedded Databases, Oracle Big Data SQL	Relational, Graph, Scalable, Structured, Client specific

[5] view from abstract concept to its realization.

based on constraint programming, we just extract this semantic information and use it in our analysis. We have not verified the validity of original data. Finally, automatic extraction of conceptual categorization from Wikipedia depends on the robustness and accuracy of its internal search engine.

4 Related Work

Leveraging textual data stored in form of natural language text, to assist in one or more phases of SDLC has been an area of interest in software engineering community [5,6,9] since a long time. Almost all such works use knowledge representation to store the extracted information [11]. If we specifically consider software architecture or design domain there have been few ontologies like [1,6] but none of these have been in terms of relating software solutions to architectural abstraction. In software development domain, although the work presented in [14] is comprehensive it is not adaptive like SDDNet. SDDNet incorporates all the changing terms and technologies and hence finds more usability within the fast growing and changing field of software development. Also, none of the ontologies, semantic networks, concept maps present for software engineering domain map the abstractions of one phase of SDLC to their realizations in other phase. Whereas, this is the basis of knowledge represented in SDDNet.

5 Conclusion and Future Work

We propose a knowledge-base - Software Design and Development Network (SDDNet)- with a semantic network at its core. It consists of entities relevant to software design and development and their inter-relationships. We have analyzed SDDNet's behavior and the results show considerable efficacy. Our final goal is to use SDDNet to drive inference engine of our automatic decision making tool for software engineering tasks.

References

1. Anvaari, M., Zimmermann, O.: Semi-automated Design Guidance Enhancer (SADGE): a framework for architectural guidance development. In: Avgeriou, P., Zdun, U. (eds.) ECSA 2014. LNCS, vol. 8627, pp. 41–49. Springer, Heidelberg (2014)
2. Bird, S., Loper, E., Klein, E.: Natural Language Processing with Python. O'Reilly Media Inc., Sebastopol (2009)
3. De Graaf, K.A., Tang, A., Liang, P., Van Vliet, H.: Ontology-based software architecture documentation. In: 2012 Joint Working IEEE/IFIP Conference on Software Architecture (WICSA) and European Conference on Software Architecture (ECSA), pp. 121–130. IEEE (2012)
4. Drieger, P.: Semantic network analysis as a method for visual text analytics. Procedia-Soc. Behav. Sci. **79**, 4–17 (2013)

5. Ilieva, M.G., Ormandjieva, O.: Automatic transition of natural language software requirements specification into formal presentation. In: Montoyo, A., Muñoz, R., Métais, E. (eds.) NLDB 2005. LNCS, vol. 3513, pp. 392–397. Springer, Heidelberg (2005)
6. López, C., Codocedo, V., Astudillo, H., Cysneiros, L.M.: Bridging the gap between software architecture rationale formalisms and actual architecture documents: an ontology-driven approach. Sci. Comput. Program. **77**(1), 66–80 (2012)
7. Miller, G.A.: Wordnet: a lexical database for english. Commun. ACM **38**(11), 39–41ʼ(1995)
8. Pressman, R.S.: Software Engineering: A Practitioner's Approach. Palgrave Macmillan, New York (2005)
9. Runeson, P., Alexandersson, M., Nyholm, O.: Detection of duplicate defect reports using natural language processing. In: 29th International Conference on Software Engineering, ICSE 2007, pp. 499–510. IEEE (2007)
10. seatgeek: fuzzywuzzy. https://github.com/seatgeek/fuzzywuzzy
11. Studer, R., Benjamins, V.R., Fensel, D.: Knowledge engineering: principles and methods. Data Knowl. Eng. **25**(1), 161–197 (1998)
12. Taylor, R.N., Medvidovic, N., Dashofy, E.M.: Software Architecture: Foundations, Theory, and Practice. Wiley Publishing, Chichester (2009)
13. Webber, J.: A programmatic introduction to Neo4j. In: Proceedings of the 3rd Annual Conference on Systems, Programming, and Applications: Software for Humanity, pp. 217–218. ACM (2012)
14. Wongthongtham, P., Chang, E., Dillon, T., Sommerville, I.: Development of a software engineering ontology for multisite software development. IEEE Trans. Knowl. Data Eng. **21**(8), 1205–1217 (2009)

Formal Concept Analysis

Parallel Attribute Exploration

Francesco Kriegel[✉]

Institute of Theoretical Computer Science,
Technische Universität Dresden, Dresden, Germany
francesco.kriegel@tu-dresden.de

Abstract. The canonical base of a formal context is a minimal set of implications that is sound and complete. A recent paper has provided a new algorithm for the parallel computation of canonical bases. An important extension is the integration of expert interaction for *Attribute Exploration* in order to explore implicational bases of inaccessible formal contexts. This paper presents and analyzes an algorithm that allows for *Parallel Attribute Exploration*.

Keywords: Formal Concept Analysis · Attribute Exploration · Canonical base · Implication · Parallel algorithm · Expert interaction · Supervised learning

1 Introduction

Implications provide an easily understandable means of logical knowledge representation. When learning terminological knowledge, it is hence straight-forward to extract valid implications from a data-set. If the data-set is complete, i.e., fully describes all individuals in the domain of interest, and if furthermore the data-set has a propositional structure or is represented as a formal context, then it suffices to compute the canonical base which is sound and complete for the set of implications holding in the data-set. This result has been found by Guigues and Duquenne [8] in the field of *Formal Concept Analysis*, where the authors utilize the notion of pseudo-intents to construct such canonical bases. A strong advantage of a canonical base is its minimality, i.e., there is no smaller set of implications which is sound and complete for the underlying data-set. Ganter [5] introduced the algorithm NextClosure for the computation of canonical bases, and proved its correctness. Previously, in the field of database theory, the deduction of dependencies has been investigated in Maier [12], but however no explicit construction of a base of dependencies was provided.

For the case of incomplete data-sets, i.e., if there are further individuals in the domain of interest that are not described in the data-set, a technique called *Attribute Exploration* has been developed by Ganter [4–6] and Stumme [13]. It allows for interaction with an expert which is able to provide unknown individuals that are not contained in the data-set but represent counterexamples to otherwise valid implications. This algorithm is merely an extension of the algorithm NextClosure by Ganter [5]. Unfortunately, the algorithm uses the lectic

© Springer International Publishing Switzerland 2016
O. Haemmerlé et al. (Eds.): ICCS 2016, LNAI 9717, pp. 91–106, 2016.
DOI: 10.1007/978-3-319-40985-6_8

order on the attribute set, which is linear, to compute the implications in the canonical base. As a consequence, it is not (obviously) possible to parallelize this default *Attribute Exploration*. However, in [10,11] we have introduced the algorithm NextClosures that is also able to compute canonical bases, but in a non-linear order which makes it possible to enumerate the elements of the canonical base in a parallel manner. In particular, the canonical base is constructed w.r.t. increasing premise cardinality. Benchmarks have shown that there is an inverse linear correlation between the computation time and the number of available CPU cores, provided that the underlying formal context is large enough, and that its performance on one CPU core is comparable to NextClosure, more specifically the quotient of the computation times for the same formal context is between $\frac{1}{3}$ and 3. The benchmarks in [10,11] should only be interpreted in a relative way – using more efficient data structures (e.g., java.util.BitSet), or faster programming languages (e.g., C++), the computation times can be decreased further. An implementation of NextClosures in the programming language Java 8 can be found in [9].

In this paper, NextClosures is extended with the possibility of expert interaction. More specifically, we assume that there is a formal context that describes the domain of interest, but it is inaccessible and there is an expert (or a set of experts) that can correctly decide whether an implication holds in this context, and if she refutes then also provides a counterexample. Additionally, there may be an observed subcontext of the full domain context, which is used to decrease the number of questions posed to the expert. Using the technique of *Attribute Exploration*, it is possible to construct a minimal implicational base of the domain context. The algorithm ParallelAttributeExploration that will be described in the following sections implements this technique and furthermore allows for a parallel execution.

Angluin et al. [1] have also investigated the problem of learning propositional Horn-theories by means of oracles. In particular, they assume that there are two experts: a *membership oracle* and an *equivalence oracle*. While the first expert decides whether a certain object satisfies the (unknown) target theory, the second expert decides whether the current theory is equivalent to the target theory (and if not, returns a counterexample). Later, Arias and Balcázar [2] have proven that this learning approach always constructs the canonical base [8] for the target theory.

This document is structured as follows. Section 2 introduces the basic notions of *Formal Concept Analysis*, and Sect. 3 defines the notion of an expert as well as provides some important statements on the interplay of formal contexts and experts. Section 4 presents the algorithm ParallelAttributeExploration, and furthermore proves its soundness and completeness. Section 5 draws a comparison with the default algorithm for *Attribute Exploration*, as well as discusses some possibilities for the integration of several experts.

2   Formal Concept Analysis

In this section we shall introduce the basic notions of *Formal Concept Analysis*, cf. [7]. A *formal context* $\mathbb{K} = (G, M, I)$ consists of a set G of *objects*, a set M of *attributes*, and an *incidence relation* $I \subseteq G \times M$ such that $g\, I\, m$ indicates that object g *has* attribute m. Furthermore, for subsets $A \subseteq G$ and $B \subseteq M$, their *derivations* are defined as follows:

$$A^I := \{\, m \in M \mid \forall g \in A\colon g\, I\, m \,\} \text{ and } B^I := \{\, g \in G \mid \forall m \in B\colon g\, I\, m \,\}.$$

It is well-known [7] that both derivation operators form a so-called *Galois connection* between the powersets $\wp(G)$ and $\wp(M)$, i.e., the following statements hold true for all subsets $A, A_1, A_2 \subseteq G$ and $B, B_1, B_2 \subseteq M$:

1. $A \subseteq B^I \Leftrightarrow B \subseteq A^I$.
2. $A \subseteq A^{II}$.
3. $A^I = A^{III}$.
4. $A_1 \subseteq A_2 \Rightarrow A_2^I \subseteq A_1^I$.

5. $B \subseteq B^{II}$.
6. $B^I = B^{III}$.
7. $B_1 \subseteq B_2 \Rightarrow B_2^I \subseteq B_1^I$.

For subsets $A \subseteq G$ and $B \subseteq M$, the pair (A, B) is a *formal concept* of $\mathbb{K}$ if $A^I = B$ and $B^I = A$. Then we refer to A as the *extent*, and to B as the *intent* of (A, B). The set of all formal concepts is denoted as $\mathfrak{B}(\mathbb{K})$, and $\mathsf{Int}(\mathbb{K})$ denotes the set of all intents of $\mathbb{K}$. $\mathfrak{B}(\mathbb{K})$ can be ordered by $(A, B) \leq (C, D)$ if $A \subseteq C$ (or dually if $B \supseteq D$), and indeed then $(\mathfrak{B}(\mathbb{K}), \leq)$ is a complete lattice where infima and suprema are given as follows:

$$\bigwedge_{t \in T}(A_t, B_t) = (\bigcap_{t \in T} A_t, (\bigcup_{t \in T} B_t)^{II}) \text{ and } \bigvee_{t \in T}(A_t, B_t) = ((\bigcup_{t \in T} A_t)^{II}, \bigcap_{t \in T} B_t).$$

An *implication* over M is an expression of the form $X \to Y$ where $X, Y \subseteq M$. The set of all implications over M is denoted by $\mathsf{Imp}(M)$. We say that $X \to Y$ is *valid* in $\mathbb{K}$, denoted as $\mathbb{K} \models X \to Y$, if $X^I \subseteq Y^I$. $\mathsf{Imp}(\mathbb{K})$ is the set of all valid implications of $\mathbb{K}$. A subset $Z \subseteq M$ is a *model* of $X \to Y$ if $X \subseteq Z$ implies $Y \subseteq Z$, denoted by $Z \models X \to Y$, and $\mathsf{Mod}(X \to Y)$ is the set of all models of $X \to Y$. Furthermore, $\mathsf{Mod}(\mathcal{L}) := \bigcap\{\, \mathsf{Mod}(X \to Y) \mid X \to Y \in \mathcal{L} \,\}$ is the set of all models of an implication set $\mathcal{L}$. It is well-known that the following statements are equivalent:

1. $X \to Y$ is valid in $\mathbb{K}$.
2. Each intent of $X \to Y$ is valid in $\mathbb{K}$.
3. Each object intent of $\mathbb{K}$ models $X \to Y$.
4. $Y \subseteq X^{II}$.

Furthermore, the relation $\models$ may be lifted to implication sets as follows: Let $\mathcal{L} \cup \{X \to Y\} \subseteq \mathsf{Imp}(M)$, then $\mathcal{L}$ *entails* $X \to Y$, symbolized by $\mathcal{L} \models X \to Y$, if every model of $\mathcal{L}$ is a model of $X \to Y$. Then $\mathsf{Imp}(\mathcal{L})$ is the set of all implications that are entailed by $\mathcal{L}$. For each subset $X \subseteq M$, there is a smallest superset $X^{\mathcal{L}}$

of X that is a model of $\mathcal{L}$, since $\mathsf{Mod}(\mathcal{L})$ is closed under intersection. It is well-known that this set can be computed as follows:

$$X^{\mathcal{L}} = \bigcup_{n \geq 1} X^{\mathcal{L}_n} \text{ where } X^{\mathcal{L}_{n+1}} := (X^{\mathcal{L}_1})^{\mathcal{L}_n} \text{ for all } n \geq 1,$$

$$\text{and } X^{\mathcal{L}_1} := X \cup \bigcup \{ Z \mid Y \to Z \in \mathcal{L} \text{ and } Y \subseteq X \}.$$

It is easy to verify that the following statements are equivalent:

1. $\mathcal{L} \models X \to Y$.
2. For all $Z \subseteq M$, $Z \models \mathcal{L}$ implies $Z \models X \to Y$.
3. For all formal contexts $\mathbb{K}$ with attribute set M, $\mathbb{K} \models \mathcal{L}$ implies $\mathbb{K} \models X \to Y$.
4. $Y \subseteq X^{\mathcal{L}}$.

Note that a formal context is just another notion for a set of propositional models (where the attributes in M are considered as propositional variables). In particular, for a formal context $\mathbb{K} = (G, M, I)$ the set $\mathcal{P}_{\mathbb{K}} := \{ \chi_{g^I}^M \mid g \in G \}$, where χ_B^M is the characteristic function of B in M, is a set of propositional models such that for each implication $X \to Y$ over M, $X \to Y$ is valid in $\mathbb{K}$ if, and only if, $\bigwedge X \to \bigwedge Y$ is valid in $\mathcal{P}_{\mathbb{K}}$.

Analogously, if $\mathcal{P}$ is a set of propositional models over a set M of propositional variables, then the formal context $\mathbb{K}_{\mathcal{P}} := (\{ p^{-1}(1) \mid p \in \mathcal{P} \}, M, \ni)$ satisfies $\mathcal{P} \models \bigwedge X \to \bigwedge Y$ if, and only if, $\mathbb{K}_{\mathcal{P}} \models X \to Y$, for all implications $X \to Y$ over M.

An *implicational base* of a formal context $\mathbb{K}$ is an implication set that is *sound*, i.e., is valid in $\mathbb{K}$, and is *complete*, i.e., entails all valid implications of $\mathbb{K}$. An implicational base is *irredundant* if none of its implications follows from the others, and is *minimal* if it has minimal cardinality among all implicational bases for $\mathbb{K}$. It is straight-forward to show that the following statements are equivalent:

1. $\mathcal{L}$ is an implicational base for $\mathbb{K}$.
2. $\mathsf{Imp}(\mathcal{L}) = \mathsf{Imp}(\mathbb{K})$.
3. $\mathsf{Mod}(\mathcal{L}) = \mathsf{Int}(\mathbb{K})$.

A *pseudo-intent* of $\mathbb{K} = (G, M, I)$ is an attribute set $P \subseteq M$ such that $P \neq P^{II}$, and $Q^{II} \subseteq P$ for all pseudo-intents $Q \subsetneq P$. The set of all pseudo-intents of $\mathbb{K}$ is denoted by $\mathsf{PsInt}(\mathbb{K})$. The *canonical base* of a formal context $\mathbb{K}$ is defined as

$$\mathcal{B}_{\mathsf{can}}(\mathbb{K}) := \{ P \to P^{II} \mid P \in \mathsf{PsInt}(\mathbb{K}) \},$$

and is a minimal implicational base for $\mathbb{K}$, cf. [5–8].

It has been shown that the set of all intents and pseudo-intents is a closure system. The corresponding closure operator $\cdot^{\mathbb{K}^*}$ is given by the following definition:

$$X^{\mathbb{K}^*} := \bigcup_{n \geq 1} X^{\mathbb{K}_n^*} \text{ where } X^{\mathbb{K}_{n+1}^*} := (X^{\mathbb{K}_1^*})^{\mathbb{K}_n^*} \text{ for all } n \geq 1,$$

$$\text{and } X^{\mathbb{K}_1^*} := X \cup \bigcup \{ P^{II} \mid P \in \mathsf{PsInt}(\mathbb{K}) \text{ and } P \subsetneq X \}.$$

More specifically, then an attribute set $X \subseteq M$ is an intent or a pseudo-intent of $\mathbb{K}$ if, and only if, $X = X^{\mathbb{K}^*}$. Additionally, we may also define a pseudo-closure operator for implication sets $\mathcal{L} \subseteq \mathsf{Imp}(M)$:

$$X^{\mathcal{L}^*} := \bigcup_{n \geq 1} X^{\mathcal{L}_n^*} \text{ where } X^{\mathcal{L}_{n+1}^*} := (X^{\mathcal{L}_1^*})^{\mathcal{L}_n^*} \text{ for all } n \geq 1,$$

$$\text{and } X^{\mathcal{L}_1^*} := X \cup \bigcup \{ Z \mid Y \to Z \in \mathcal{L} \text{ and } Y \subsetneq X \}.$$

It is readily verified that both closure operators $\cdot^{\mathbb{K}^*}$ and $\cdot^{\mathcal{L}^*}$ coincide in case $\mathcal{L} = \mathcal{B}_{\mathsf{can}}(\mathbb{K})$.

3   Experts

An *expert* is an oracle that correctly answers questions in a certain domain of interest. For our purposes, the questions are expressed in form of implications, and an expert may either *accept* or *decline*. If the expert accepts an implication, then it must hold for all objects in the domain of interest, and otherwise she must return a refutation, i.e., an object that serves as a counterexample. In this section, we will formally define the notion of an expert, and provide some basic statements.

Definition 1 (Expert, [3, Definition 6.1.2]). *Let M be a set of attributes. An* expert *on M is a partial mapping $\chi \colon \mathsf{Imp}(M) \to_{\mathsf{p}} \wp(M)$ that satisfies the following properties:*

1. *If $\chi(X \to Y)$ is defined, then the value is not a model of $X \to Y$, i.e., $\chi(X \to Y) = C$ implies $X \subseteq C$ and $Y \not\subseteq C$. Furthermore, we then call C a counterexample* against $X \to Y$.
   (Experts return counterexamples for refuted implications.)
2. *If $\chi(X \to Y)$ is undefined, then every other counterexample given by χ must be a model of $X \to Y$, i.e., $\chi(U \to V) = C$ implies $X \not\subseteq C$ or $Y \subseteq C$.*
   (Counterexamples do not refute accepted implications.)

Furthermore, we say that χ accepts $X \to Y$, and denote this as $\chi \models X \to Y$, if $\chi(X \to Y)$ is undefined, and that χ refutes $X \to Y$ otherwise. The set of all accepted implications of χ is denoted by $\mathsf{Imp}(\chi)$, and the set of all counterexamples of χ is denoted by $\mathsf{Cex}(\chi) := \{ C \mid \exists X, Y \subseteq M \colon \chi(X \to Y) = C \}$.

There is a correspondence between formal contexts and experts as follows:

Definition 2 (Induced Expert). *An expert χ on M is induced by a formal context $\mathbb{K} = (G, M, I)$ if it accepts exactly those implications that are valid in $\mathbb{K}$, i.e., $\mathsf{Imp}(\mathbb{K}) = \mathsf{Imp}(\chi)$. If χ is an expert on M, then its induced formal context is $\mathbb{K}_\chi := (\mathsf{Cex}(\chi), M, \ni)$.*

Lemma 3. *Let $\mathbb{K} = (G, M, I)$ be a formal context and χ an expert on M. Then χ is induced by $\mathbb{K}$ if, and only if, it accepts only valid implications of $\mathbb{K}$, and all counterexamples are intents of $\mathbb{K}$, i.e., $\mathsf{Imp}(\chi) \subseteq \mathsf{Imp}(\mathbb{K})$ as well as $\mathsf{Cex}(\chi) \subseteq \mathsf{Int}(\mathbb{K})$.*

Proof. The if-direction is trivial. For the converse direction, assume that χ accepts exactly those implications that are valid in $\mathbb{K}$. Of course, then each implication accepted by χ is valid in $\mathbb{K}$. Assume that χ refutes an implication with a counterexample C. Since the implication $C \to C^{II}$ is trivially valid in $\mathbb{K}$, χ accepts $C \to C^{II}$. Consequently, the counterexample C must be a model of $C \to C^{II}$, i.e., C is an intent of $\mathbb{K}$.   □

Lemma 4. *If χ is an expert on M, then χ is an induced expert of $\mathbb{K}_\chi$.*

Proof. Let χ be an expert on M, and consider an implication $X \to Y$. If χ accepts $X \to Y$, then by Statement 2 of Definition 1 all counterexamples of χ are models of $X \to Y$. We conclude that all object intents of the $\mathbb{K}_\chi$ are models of $X \to Y$, i.e., $X \to Y$ is valid in $\mathbb{K}_\chi$.

Vice versa, if $X \to Y$ is valid in $\mathbb{K}_\chi$, then all intents of $\mathbb{K}_\chi$ are models of $X \to Y$. Hence, χ cannot refute $X \to Y$, as the counterexample would be an object intent of $\mathbb{K}_\chi$, but would not be a model of $X \to Y$.   □

Corollary 5. *If $\mathbb{K}$ is a formal context with an induced expert χ, then an implication is valid in $\mathbb{K}$ if, and only if, it is valid in $\mathbb{K}_\chi$. Furthermore, then every (minimal) implicational base of $\mathbb{K}$ is a (minimal) implicational base of $\mathbb{K}_\chi$, and vice versa, i.e., the sets of intents of $\mathbb{K}$ and $\mathbb{K}_\chi$ coincide.*

Definition 6 (Optimal Expert). *An expert χ is optimal if for all implications $X \to Y$, it is true that χ accepts $X \to Y \cap C$ if χ refutes $X \to Y$ with counterexample C.*

Lemma 7. *Let $\mathbb{K}$ be a formal context. An induced expert χ of $\mathbb{K}$ is optimal if, and only if, for all implications $X \to Y$, $\chi(X \to Y) = C$ implies $Y \cap C \subseteq X^{II} \subseteq C$.*

Furthermore, the canonical expert $\chi_\mathbb{K}$ for $\mathbb{K}$ is an optimal induced expert for $\mathbb{K}$, where

$$\chi_\mathbb{K}(X \to Y) := \begin{cases} undefined & if\ \mathbb{K} \models X \to Y, \\ X^{II} & otherwise. \end{cases}$$

Proof. The if-direction is obvious. Vice versa, let χ be optimal for $\mathbb{K}$, and consider an implication $X \to Y$ that is refuted by χ with counterexample C, i.e., $X \subseteq C$ and $Y \not\subseteq C$. Since χ is induced by $\mathbb{K}$, C is an intent, and so $X^{II} \subseteq C$. Furthermore, as χ is optimal, $X \to Y \cap C$ is valid in $\mathbb{K}$, i.e., $Y \cap C \subseteq X^{II}$.

Eventually, $\chi_\mathbb{K}$ is an induced expert for $\mathbb{K}$, since it accepts all implications that are valid in $\mathbb{K}$, and all counterexamples are intents of $\mathbb{K}$. Furthermore, it is optimal, as implications $X \to Y \cap X^{II}$ are trivially valid in $\mathbb{K}$.   □

Lemma 8. *Let χ be an expert. Then there is an optimal expert $\widehat{\chi}$ such that both accept the same implications, i.e., $\mathsf{Imp}(\chi) = \mathsf{Imp}(\widehat{\chi})$.*

Proof. Consider an expert χ. We construct an equivalent optimal expert $\widehat{\chi}$ as follows. Let $X \to Y$ be an arbitrary implication. If χ accepts $X \to Y$, then $\widehat{\chi}$ accepts $X \to Y$, too. If χ rejects $X \to Y$, then proceed in the following way. Let

$Y_0 := Y$, and $n := 0$. While χ rejects $X \to Y_n$ with the counterexample Z_n, set $Y_{n+1} := Y_n \cap Z_n$ and increase n. Eventually, define $\widehat{\chi}(X \to Y) := \bigcap_{k=0}^{n} Z_k$.

It remains to prove that $\widehat{\chi}$ is optimal and accepts the same implications as χ. Assume that $\widehat{\chi}$ refutes $X \to Y$ with counterexample Z. Then there exists a sequence $Z_0, \ldots, Z_n$ of counterexamples of χ as above such that Z equals their intersection, and χ accepts $X \to Y \cap Z$. By construction, then also $\widehat{\chi}$ accepts the adjusted implication $X \to Y \cap Z$.

By definition, we already know that $\mathsf{Imp}(\chi) \subseteq \mathsf{Imp}(\widehat{\chi})$. Vice versa, since χ rejects only if $\widehat{\chi}$ rejects, we conclude that χ accepts if $\widehat{\chi}$ accepts. $\square$

Proposition 9. *Let $\mathbb{K} = (G, M, I)$ be a formal context and χ an expert on M. Then the following statements are equivalent:*

1. *χ is induced by $\mathbb{K}$.*
2. *χ accepts exactly those implications that are valid in $\mathbb{K}$, i.e., $\mathsf{Imp}(\chi) = \mathsf{Imp}(\mathbb{K})$.*
3. *χ accepts only valid implications of $\mathbb{K}$, and all counterexamples are intents of $\mathbb{K}$, i.e., $\mathsf{Imp}(\chi) \subseteq \mathsf{Imp}(\mathbb{K})$ and $\mathsf{Cex}(\chi) \subseteq \mathsf{Int}(\mathbb{K})$.*
4. *Each intent of $\mathbb{K}$ is an intersection of counterexamples of χ, and all counterexamples are intents of $\mathbb{K}$, i.e., $\langle \mathsf{Cex}(\chi) \rangle_{\cap} = \mathsf{Int}(\mathbb{K})$.*

Proof. Statements 1 to 3 are equivalent by Definition 2 and Lemma 3.

3.$\Leftrightarrow$4. We consider the *optimization* $\widehat{\chi}$ from Lemma 8, then by construction it is true that every counterexample of $\widehat{\chi}$ is an intersection of counterexamples of χ. The maximal intent M is obtained as the empty intersection (of counterexamples). For each intent $B = B^{II}$ where $B \neq M$, the implication $B \to M$ is invalid in $\mathbb{K}$, and thus must be rejected by $\widehat{\chi}$ with a counterexample C. Then Lemma 7 yields $C = M \cap C \subseteq B^{II} \subseteq C$.

Vice versa, let every intent of $\mathbb{K}$ be an intersection of counterexamples of χ, and assume that all counterexamples are intents of $\mathbb{K}$. Consider an implication $X \to Y$ that is not valid in $\mathbb{K}$, i.e., $Y \not\subseteq X^{II}$. In particular, then X^{II} is an intersection of counterexamples $C_1, \ldots, C_n$ of χ, and the C_i are intents of $\mathbb{K}$. If χ accepts $X \to Y$, then Statement 2 of Definition 1 implies that all counterexamples of χ are models of $X \to Y$, and in particular each C_i is a model of $X \to Y$. Since the set of models of an implication is closed under intersection, X^{II} must be a model of $X \to Y$. Contradiction! $\square$

Lemma 10. *Let χ be an induced expert of a formal context $\mathbb{K}$. If χ is optimal, then $\mathsf{Cex}(\chi)$ is closed under non-empty intersections.*

Proof. Let χ be optimal, and consider two counterexamples C_1 and C_2. The implication $C_1 \cap C_2 \to M$ must be rejected by χ with a counterexample C such that $M \cap C \subseteq (C_1 \cap C_2)^{II} \subseteq C$, i.e., $C = C_1 \cap C_2$, since both C_i are intents. $\square$

However, the converse statement does not hold in general. To see this, consider a formal context $\mathbb{K}$ over $M := \{a, b, c\}$ where $\mathsf{Int}(\mathbb{K}) = \{\{a\}, \{a, b\}, \{a, b, c\}\}$. Hence, for each induced expert χ, we have $\emptyset \neq \mathsf{Cex}(\chi) \subseteq \{\{a\}, \{a, b\}\}$ and it is easily verified that for each choice, the set of counterexamples is closed under non-empty intersections. However, an induced expert χ with $\chi(\{a\} \to \{b, c\}) := \{a, b\}$ is not optimal.

4   Parallel Attribute Exploration

As the next step, we will introduce the algorithm for *Parallel Attribute Exploration*. Assume that we want to compute a minimal implicational base for an inaccessible formal context $\mathbb{D}$ with attribute set M, and we have observed an induced subcontext $\mathbb{K} = (G, M, I)$ of $\mathbb{D}$ as well as we know an expert χ that is induced by $\mathbb{D}$. Of course, it is not useful to simply compute an implicational base of $\mathbb{K}$, as wrong conclusions could be drawn. There are two naïve ways to accomplish the computation of a base. According to Corollary 5, we may construct the formal context $\mathbb{K}_\chi$ induced by χ, and compute its canonical base, e.g., by means of the algorithms in [5,6,11]. However, this is certainly no practical approach, as its puts a high workload on the expert by posing all possible implications as questions to her. A slight improvement would consist in first checking whether the implication in question is already refuted by the known subcontext $\mathbb{K}$, and only in case of validity ask the expert for acceptance. Of course, all implications holding in $\mathbb{D}$ are valid in $\mathbb{K}$, too. Unfortunately, this modification is still not efficient, since the number of questions posed to the expert will not be minimal in order to compute an implicational base. In general, calls to the expert are expensive, and it should be ensured that only a minimal amount of work is put on her. This is the starting point for an algorithm called *Attribute Exploration*, which is basically an extension of `NextClosure` with expert interaction as introduced by Ganter [5,6] and Ganter and Wille [7]. It enumerates all pseudo-intents of the context $\mathbb{D}$ and only poses implicational questions to the expert whose premise is a pseudo-intent. This ensures the minimality on the number of questions w.r.t. both $\mathbb{K}$ and χ. If furthermore χ is optimal, then the number of questions is minimal w.r.t. $\mathbb{K}$, i.e., for each other expert χ' induced by $\mathbb{D}$, χ' must answer at least as many questions as χ.

While *Attribute Exploration* [5–7] constructs the canonical base of $\mathbb{D}$ in a lectic order, Algorithm 1 computes it w.r.t. increasing premise cardinality, which in turn allows to process all implications with the same premise cardinality in parallel. Note that Algorithm 1 is an extension of [11, Algorithm 1] with expert interaction.

Definition 11. *Let (G_1, M, I_1) and (G_2, M, I_2) be two formal contexts with disjoint object sets and the same attribute set. Their* subposition *is defined as the formal context*

$$\frac{(G_1, M, I_1)}{(G_2, M, I_2)} := (G_1 \cup G_2, M, I_1 \cup I_2).$$

For a formal context $\mathbb{K} = (G, M, I)$ and an attribute set $X \subseteq M$, the formal context

$$\mathbb{K}[X] := (G \cup \{g_X\}, M, I \cup \{g_X\} \times X)$$

is obtained by adding a new object $g_X \notin G$ that has all attributes from X. In particular, $\mathbb{K}[X]$ is a subposition of $\mathbb{K}$ and the row X. For a sequence $X_1, X_2, \ldots, X_n \subseteq M$, we inductively define

$$\mathbb{K}[X_1, X_2, \ldots, X_n] := (\mathbb{K}[X_1])[X_2, \ldots, X_n].$$

Furthermore, we write $\mathbb{K} \leq_M \mathbb{D}$ if there is a context $\mathbb{U}$ such that $\mathbb{D} = \frac{\mathbb{K}}{\mathbb{U}}$.

If $\mathcal{L}$ is an implication set, and $k \in \mathbb{N}$, then $\mathcal{L}\lceil_k := \{ X \to Y \in \mathcal{L} \mid |X| \le k \}$ contains all implications from $\mathcal{L}$ whose premises have a cardinality of at most k. Furthermore, we define $\mathsf{PsInt}(\mathbb{K})\lceil_k := \{ P \in \mathsf{PsInt}(\mathbb{K}) \mid |P| \le k \}$

Lemma 12. *Let* $\mathbb{K} = (G, M, I)$ *be a formal context, and* $X \subseteq M$ *an attribute set. Furthermore, denote the incidence relation of* $\mathbb{K}[X]$ *by* J. *Then the following statements hold:*

1. *For all attribute sets* $B \subseteq M$, *it is true that*

$$B^{JJ} = \begin{cases} B^{II} \cap X & \text{if } B \subseteq X, \text{ and} \\ B^{II} & \text{otherwise.} \end{cases}$$

2. *If* X *is a model of the implication* $Y \to Y^{II}$, *then* $Y^{II} = Y^{JJ}$.
3. *If* X *is a model of all implications* $P \to P^{II}$ *where* P *is a pseudo-intent of* $\mathbb{K}$ *with* $|P| \le k$, *then the pseudo-intents of* $\mathbb{K}$ *and* $\mathbb{K}[X]$ *with cardinality* $\le k$ *coincide, i.e.,* $X \models \mathcal{B}_{\mathsf{can}}(\mathbb{K})\lceil_k$ *implies* $\mathsf{PsInt}(\mathbb{K})\lceil_k = \mathsf{PsInt}(\mathbb{K}[X])\lceil_k$.

Proof. 1. Let $B \subseteq X$, then $B^J = B^I \cup \{g_X\}$, and hence $B^{JJ} = (B^I \cup \{g_X\})^J = B^{IJ} \cap g_X^J = B^{II} \cap X$. Otherwise, $B^J = B^I$, and thus $B^{JJ} = B^{IJ} = B^{II}$.

2. Assume that $X \models Y \to Y^{II}$, i.e., $Y \subseteq X$ implies $Y^{II} \subseteq X$. If $Y \subseteq X$, then $Y^{JJ} = Y^{II} \cap X = Y^{II}$. Otherwise, $Y^{JJ} = Y^{II}$ follows directly.

3. We prove the statement by induction on k. First let $k = 0$. Obviously, $\emptyset$ is the only set of cardinality 0. Since it has no strict subsets, it is a pseudo-intent if, and only if, it is no intent. If $\emptyset$ is a pseudo-intent of $\mathbb{K}$, then X is a model of $\emptyset \to \emptyset^{II}$. Statement 2 yields $\emptyset^{II} = \emptyset^{JJ}$, and thus $\emptyset \ne \emptyset^{JJ}$. If otherwise $\emptyset$ is an intent of $\mathbb{K}$, then it holds that $\emptyset = \emptyset^{II} \supseteq \emptyset^{II} \cap X \supseteq \emptyset$, i.e., $\emptyset$ must be an intent of $\mathbb{K}[X]$, too.

Now assume that the induction hypothesis holds for k. Consider a pseudo-intent P of $\mathbb{K}$ with $|P| = k + 1$. Since X is a model of $P \to P^{II}$, Statement 2 yields $P^{II} = P^{JJ}$, and hence P is no intent of $\mathbb{K}[X]$. Now let $Q \subsetneq P$ be a pseudo-intent of $\mathbb{K}[X]$. Then $|Q| \le k$, and hence Q is a pseudo-intent of $\mathbb{K}$ by induction hypothesis. Consequently, $Q^{II} \subseteq P$, and thus $Q^{JJ} \subseteq P$.

Vice versa, let P be a pseudo-intent of $\mathbb{K}[X]$ with $|P| = k + 1$. Then P is no intent of $\mathbb{K}$, as $P \ne P^{JJ} \subseteq P^{II}$. Consider a pseudo-intent $Q \subsetneq P$ of $\mathbb{K}$. Then Q must be a $\mathbb{K}[X]$-pseudo-intent by induction hypothesis. Furthermore, $Q^{JJ} \subseteq P$. Since X is a model of $Q \to Q^{II}$, Statement 2 implies $Q^{JJ} = Q^{II}$, and thus $Q^{II} \subseteq P$. $\square$

As an immediate consequence we deduce from the preceding lemma, more specifically from Statements 2 and 3, that the following corollary holds.

Corollary 13. *Let* $\mathbb{K} = (G, M, I)$ *be a formal context, and* $X \subseteq M$ *an attribute set. If* X *is a model of* $\mathcal{B}_{\mathsf{can}}(\mathbb{K})\lceil_k$, *then* $\mathcal{B}_{\mathsf{can}}(\mathbb{K})\lceil_k = \mathcal{B}_{\mathsf{can}}(\mathbb{K}[X])\lceil_k$.

By successive application of the previous corollary we get the following statement.

Algorithm 1. `ParallelAttributeExploration`

Require: a formal context $\mathbb{K} = (G, M, I)$
Require: an expert χ on M
1 $\mathbf{C} := \{\emptyset\}$, $\mathcal{L} := \emptyset$
2 **for** $k = 0, 1, \ldots, |M| - 1$ **do**
3   **for all** $C \in \mathbf{C}$ with $|C| = k$ **do in parallel**
4   **if** $C = C^{\mathcal{L}^*}$ **then**
5   **while** $C \neq C^{II}$ **and** $\chi(C \to C^{II}) = X$ **do**
6          $\mathbb{K} := \mathbb{K}[X]$
7   **if** $C \neq C^{II}$ **then**
8          $\mathcal{L} := \mathcal{L} \cup \{C \to C^{II}\}$
9          $\mathbf{C} := \mathbf{C} \cup \{C^{II} \cup \{m\} \mid m \notin C^{II}\}$
10   **else**
11       $\mathbf{C} := \mathbf{C} \cup \{C^{\mathcal{L}^*}\}$
12   Wait for termination of all parallel processes.
13 **return** $(\mathbb{K}, \mathcal{L})$

Lemma 14. *Let* $\mathbb{K} = (G, M, I)$ *be a formal context, and* $X_1, \ldots, X_n \subseteq M$ *attribute sets. If each* X_i *is a model of* $\mathcal{B}_{\mathsf{can}}(\mathbb{K})\!\restriction_k$, *then* $\mathcal{B}_{\mathsf{can}}(\mathbb{K})\!\restriction_k = \mathcal{B}_{\mathsf{can}}(\mathbb{K}[X_1, \ldots, X_n])\!\restriction_k$.

Proof. We show by induction on $i \in \{1, \ldots, n\}$ that $\mathcal{B}_{\mathsf{can}}(\mathbb{K})\!\restriction_k = \mathcal{B}_{\mathsf{can}}(\mathbb{K}[X_1, \ldots, X_i])\!\restriction_k$. The induction base follows from Corollary 13. Now assume that $i < n$ and the induction hypothesis holds for i. Then $X_{i+1} \models \mathcal{B}_{\mathsf{can}}(\mathbb{K})\!\restriction_k = \mathcal{B}_{\mathsf{can}}(\mathbb{K}[X_1, \ldots, X_i])\!\restriction_k$, and again by Corollary 13 we conclude $\mathcal{B}_{\mathsf{can}}(\mathbb{K}[X_1, \ldots, X_i])\!\restriction_k = \mathcal{B}_{\mathsf{can}}(\mathbb{K}[X_1, \ldots, X_i, X_{i+1}])\!\restriction_k$. □

In [11] we have shown that in order to correctly determine whether an attribute set with at most k elements is an intent or pseudo-intent of $\mathbb{K}$, it suffices to know the part of the canonical base that consists of all implications whose premise has a cardinality smaller than k. More specifically, we cite the following corollary.

Corollary 15 ([11, Corollary 3]). *If* $\mathcal{L}$ *contains all implications* $P \to P^{II}$ *where* P *is a pseudo-intent of* $\mathbb{K}$ *with* $|P| < k$, *and otherwise only implications with premise cardinality* k, *then for all attribute sets* $X \subseteq M$ *with* $|X| \leq k$ *the following statements are equivalent:*

1. X *is an intent or a pseudo-intent of* $\mathbb{K}$.
2. X *is* $\mathcal{L}^*$-*closed.*

Algorithm 1 describes *Parallel Attribute Exploration* in pseudo-code. If the expert χ is optimal, then the **while**-statement in Line 5 may be replaced with the analogous **if**-statement, since after χ refutes an implication $C \to C^{II}$ with a counterexample X we have that χ accepts $C \to C^{II} \cap X$, and $C^{JJ} = C^{II} \cap X$ where J is the incidence relation of $\mathbb{K}[X]$, i.e., in the second iteration, the condition of the **while**-statement always evaluates to **false**. In particular, the optimality of the expert is no restriction, since according to Lemma 8 we may always *optimize* an expert.

In the following text we will analyze Algorithm 1, and show its soundness, completeness, and termination. Beforehand, we define the following notions:

1. A run of `ParallelAttributeExploration` is *in state k* if all candidates of cardinality k have been processed, but none of cardinality $k + 1$.
2. $\mathbf{C}_k$ denotes the set of candidates in state k.
3. $\mathcal{L}_k$ denotes the set of implications in state k.
4. $\mathbb{K}_k := (G_k, M, I_k)$ denotes the formal context in state k.
5. $X_k^1, \ldots, X_k^{n_k}$ denote all counterexamples provided by the expert between states k and $k + 1$, i.e., it is true that $\mathbb{K}_k[X_k^1, \ldots, X_k^{n_k}] = \mathbb{K}_{k+1}$.

Proposition 16. *Let $\mathbb{K} = (G, M, I)$ be a formal context, and χ an expert on M, such that all implications accepted by χ are valid in $\mathbb{K}$. Further assume that Algorithm 1 is started on $(\mathbb{K}, \chi)$ as input, and is currently in state k. Then the following statements are satisfied:*

1. *$\mathbf{C}_k$ contains all pseudo-intents of $\mathbb{K}_{k+1}$ with cardinality $k + 1$.*
2. *$\mathcal{L}_k$ consists of all implications $P \to P^{I_k I_k}$ where P is a pseudo-intent of $\mathbb{K}_k$ with cardinality $\leq k$, i.e., $\mathcal{B}_{\mathsf{can}}(\mathbb{K}_k)\!\restriction_k = \mathcal{L}_k$.*
3. *Between the states k and $k + 1$, every attribute set with cardinality $k + 1$ is $\mathcal{L}^*$-closed if, and only if, it is either an intent or a pseudo-intent of $\mathbb{K}_{k+1}$.*

Proof. W.l.o.g. assume that the expert χ is optimal, and Line 5 of Algorithm 1 has been replaced with the analogous `if`-statement as discussed above.

We show the statements by induction on k. For the base case assume $k = -1$, as the initial state is -1. The candidate set is initialized as $\{\emptyset\}$, and thus $\mathbf{C}_{-1}$ indeed contains all pseudo-intents with 0 elements. As there are no pseudo-intents of $\mathbb{K}_{-1} = \mathbb{K}$ with at most -1 elements, the initial implication set $\mathcal{L}_{-1} = \emptyset$ satisfies Statement 2. Between the states -1 and 0 all candidates with 0 elements are processed, i.e., only $\emptyset$ is processed. Obviously, $\emptyset$ is either an intent or a pseudo-intent of $\mathbb{K}_0$, i.e., it is $(\mathbb{K}_0)^*$-closed. Furthermore, $\emptyset$ has no strict subsets, and hence it must be $\mathcal{L}^*$-closed for all implication sets $\mathcal{L}$ between states -1 and 0, i.e., $\mathcal{L}_{-1} \subseteq \mathcal{L} \subseteq \mathcal{L}_0$.

For the induction step assume that the statements hold for all states $\leq k$.

2. We will prove that $\mathcal{B}_{\mathsf{can}}(\mathbb{K}_{k+1})\!\restriction_{k+1} = \mathcal{L}_{k+1}$. Statement 2 of the induction hypothesis yields that $\mathcal{B}_{\mathsf{can}}(\mathbb{K}_k)\!\restriction_k = \mathcal{L}_k$. All counterexamples provided by the expert between states k and $k + 1$ are models of $\mathcal{L}_k$, as the expert has accepted all implications in $\mathcal{L}_k$. As a consequence, Lemma 14 implies $\mathcal{B}_{\mathsf{can}}(\mathbb{K}_k)\!\restriction_k = \mathcal{B}_{\mathsf{can}}(\mathbb{K}_{k+1})\!\restriction_k$. Since Algorithm 1 does not remove or modify any implications in $\mathcal{L}$, and between the states k and $k + 1$ only implications with a premise cardinality of $k + 1$ are added to $\mathcal{L}$, it is true that $\mathcal{L}_{k+1}\!\restriction_k = \mathcal{L}_k$ and $\mathcal{L}_{k+1}$ cannot contain any implications with a premise cardinality $> k + 1$. Hence, $\mathcal{L}_{k+1}$ contains $\mathcal{B}_{\mathsf{can}}(\mathbb{K}_{k+1})\!\restriction_k$, and it remains to show that $\mathcal{L}_{k+1}$ contains all implications of $\mathcal{B}_{\mathsf{can}}(\mathbb{K}_{k+1})$ with premise cardinality $k + 1$.

By Statement 1 of the induction hypothesis, the candidate set $\mathbf{C}_k$ contains all pseudo-intents of $\mathbb{K}_{k+1}$ with $k+1$ elements. Of course, all these candidates are processed in Lines 4–12 of Algorithm 1 between the states k and $k+1$. Then Statement 3 of the induction hypothesis yields that for each candidate C between the states k and $k+1$, C is $\mathcal{L}^*$-closed if, and only if, C is an intent or a pseudo-intent of $\mathbb{K}_{k+1}$. Consequently, each pseudo-intent of $\mathbb{K}_{k+1}$ of cardinality $k+1$ is recognized in Line 4. Now consider one such recognized pseudo-intent C. If it were an intent of the current formal context $\mathbb{K}$, then also of $\mathbb{K}_{k+1}$. Thus, the test for non-closedness in Line 5 passes, and the question $C \to C^{II}$ is posed to the expert χ in Line 5. If the conclusion C^{II} is too large, i.e., χ rejects the implication, then the returned counterexample X is added as a new row to $\mathbb{K}$ in Line 6. After execution of Lines 5 and 6, the implication $C \to C^{II}$ is accepted, and hence is valid in $\mathbb{K}_{k+1}$. All other counterexamples provided by the expert between states k and $k+1$ must be models of $C \to C^{II}$, and by a repeated application of Statement 2 of Lemma 12 we conclude that $C^{II} = C^{I_{k+1}I_{k+1}}$. It follows that $C \to C^{II}$ is indeed an implication of the canonical base of $\mathbb{K}_{k+1}$, and it is contained in $\mathcal{L}_{k+1}$, since it has been added to $\mathcal{L}$ in Line 8.

Eventually, it remains to show that there are no other implications in $\mathcal{L}_{k+1}$ with premise cardinality $k+1$ which are not in the canonical base of $\mathbb{K}_{k+1}$. Consider any candidate C between states k and $k+1$ that is no pseudo-intent of $\mathbb{K}_{k+1}$. An implication with premise C could only have been added to $\mathcal{L}$ if C is recognized as $\mathcal{L}^*$-closed in Line 4, i.e., only if C is an intent of $\mathbb{K}_{k+1}$. If C is also an intent of the current context $\mathbb{K}$, then no implication with premise C is added to $\mathcal{L}$, cf. Lines 4–8. Otherwise, if C is no intent of the current context $\mathbb{K}$, then the question $C \to C^{II}$ must be rejected by χ with a counterexample X (that is trivially an intent of $\mathbb{K}[X]$). Furthermore, then it holds that $C^{JJ} = C^{II} \cap X$ where J is the incidence relation of $\mathbb{K}[X]$. It remains to prove that $C = C^{JJ}$. The adjusted implication $C \to C^{II} \cap X$ is valid in $\mathbb{K}_{k+1}$, as it trivially holds in the current context $\mathbb{K}$ and the expert must accept it due to optimality, i.e., all counterexamples provided by χ (between states k and $k+1$) are models of the implication. Consequently, $C^{II} \cap X$ is a subset of $C^{I_{k+1}I_{k+1}} = C$, and thus $C = C^{JJ}$. It follows that the check for non-closedness in Line 7 fails, and hence no implication with premise C is added to $\mathcal{L}$ in Line 8.

3. We have already shown that $\mathcal{B}_{\mathsf{can}}(\mathbb{K}_{k+1})\!\restriction_{k+1} = \mathcal{L}_{k+1}$. Lemma 14 states that $\mathcal{B}_{\mathsf{can}}(\mathbb{K}_{k+1})\!\restriction_{k+1} = \mathcal{B}_{\mathsf{can}}(\mathbb{K}_{k+2})\!\restriction_{k+1}$, since all counterexamples $X_{k+1}^1, \ldots, X_{k+1}^{n_{k+1}}$ are models of $\mathcal{L}_{k+1}$. Consequently, for each implication set $\mathcal{L}$ with $\mathcal{L}_{k+1} \subseteq \mathcal{L} \subseteq \mathcal{L}_{k+2}$, Corollary 15 yields that an attribute set of cardinality $k+2$ is an intent or a pseudo-intent of $\mathbb{K}_{k+2}$ if, and only if, it is $\mathcal{L}^*$-closed, since $\mathcal{L}$ is a superset of $\mathcal{B}_{\mathsf{can}}(\mathbb{K}_{k+2})\!\restriction_{k+1}$ and furthermore only contains implications with a premise cardinality $k+2$.

1. Let P be a pseudo-intent of $\mathbb{K}_{k+2}$ with cardinality $k+2$. We have to show that P occurs as a candidate in $\mathbf{C}_{k+1}$. Beforehand, we prove an auxiliary lemma:

Lemma 17. *If $\ell < k+2$, then for all h with $\ell \leq h \leq k+2$ it holds that* $\mathsf{PsInt}(\mathbb{K}_\ell)\!\restriction_\ell = \mathsf{PsInt}(\mathbb{K}_h)\!\restriction_\ell$ *and* $\mathcal{B}_{\mathsf{can}}(\mathbb{K}_\ell)\!\restriction_\ell = \mathcal{B}_{\mathsf{can}}(\mathbb{K}_h)\!\restriction_\ell$.

Proof. Assume that $\ell < k + 2$. We prove the claim by induction on h. The base case $h = \ell$ is trivial. For the inductive step assume that the statement holds for h with $\ell \leq h < k+2$. In particular, then $\mathcal{L}_h = \mathcal{B}_{\mathsf{can}}(\mathbb{K}_h)\lceil_h$. We proceed by showing the inner induction: $\mathcal{B}_{\mathsf{can}}(\mathbb{K}_h)\lceil_\ell = \mathcal{B}_{\mathsf{can}}(\mathbb{K}_h[X_h^1, \ldots, X_h^i])\lceil_\ell$ for all $i \in \{1, \ldots, n_h\}$.

base case: $X_h^1 \models \mathcal{L}_h = \mathcal{B}_{\mathsf{can}}(\mathbb{K}_h)\lceil_h \supseteq \mathcal{B}_{\mathsf{can}}(\mathbb{K}_h)\lceil_\ell$ and thus Corollary 15 yields that $\mathcal{B}_{\mathsf{can}}(\mathbb{K}_h)\lceil_\ell = \mathcal{B}_{\mathsf{can}}(\mathbb{K}_h[X_h^1])\lceil_\ell$.

inductive step: $X_h^i \models \mathcal{L}_h = \mathcal{B}_{\mathsf{can}}(\mathbb{K}_h)\lceil_h \supseteq \mathcal{B}_{\mathsf{can}}(\mathbb{K}_h)\lceil_\ell = \mathcal{B}_{\mathsf{can}}(\mathbb{K}_h[X_h^1, \ldots, X_h^{i-1}])\lceil_\ell$ and hence Corollary 15 implies $\mathcal{B}_{\mathsf{can}}(\mathbb{K}_h)\lceil_\ell = \mathcal{B}_{\mathsf{can}}(\mathbb{K}_h[X_h^1, \ldots, X_h^i])\lceil_\ell$.

Since the statement holds in particular for $i = n_h$, we conclude that $\mathcal{B}_{\mathsf{can}}(\mathbb{K}_h)\lceil_\ell = \mathcal{B}_{\mathsf{can}}(\mathbb{K}_{h+1})\lceil_\ell$, since $\mathbb{K}_{h+1} = \mathbb{K}_h[X_h^1, \ldots, X_h^{n_h}]$. □

Assume that there is a pseudo-intent Q of $\mathbb{K}_{k+2}$ that is maximal w.r.t. $Q \subsetneq P$. Then it holds that $Q \subsetneq Q^{I_{k+2}I_{k+2}} \subsetneq P$, and $Q^{I_{k+2}I_{k+2}}$ is the only intent of $\mathbb{K}_{k+2}$ between Q and P. Let $\ell := |Q|$, i.e., $Q \in \mathsf{PsInt}(\mathbb{K}_{k+2})\lceil_\ell$. Then Q must be a pseudo-intent of $\mathbb{K}_\ell$, cf. Lemma 17. Consequently, $\mathcal{L}_\ell$ contains $Q \rightarrow Q^{I_\ell I_\ell}$, it is true that $Q^{I_\ell I_\ell} = Q^{I_{k+2}I_{k+2}}$, and the candidates $Q^{I_\ell I_\ell} \cup \{m\}$ for $m \in P \setminus Q^{I_\ell I_\ell}$ have been added to $\mathbf{C}$, cf. Line 9. Hence, define the sequence

$$C_0 := Q^{I_\ell I_\ell} \cup \{m\} \quad \text{where } m \in P \setminus Q^{I_\ell I_\ell}, \text{ and}$$
$$C_{i+1} := (C_i)^{\mathcal{L}^*} \quad \text{where } \mathcal{L}_{|C_i|-1} \subseteq \mathcal{L} \subseteq \mathcal{L}_{|C_i|}.$$

The attribute m for the first element C_0 of the sequence may be chosen arbitrarily. Furthermore, all following elements are well-defined, since implications in $\mathcal{L}_{|C_i|} \setminus \mathcal{L}_{|C_i|-1}$ have no influence on the closure of C_i. It is obvious that each C_i occurs as a candidate during the algorithm's run, cf. Lines 9 and 11, and that the sequence increases, i.e., $C_i \subseteq C_{i+1}$ for all indices i. We now prove by induction on i that $C_i \subseteq P$. The base case for $i = 0$ is trivial. Assume that $C_i \subseteq P$. Consider any implication set $\mathcal{L}$ where $\mathcal{L}_{|C_i|-1} \subseteq \mathcal{L} \subseteq \mathcal{L}_{|C_i|}$. Then $C_{i+1} = (C_i)^{\mathcal{L}^*}$. Furthermore, we have that $|C_i| \leq k + 2$, and thus $\mathcal{L}_{|C_i|} \subseteq \mathcal{L}_{k+2}$. Consequently,

$$C_{i+1} = (C_i)^{\mathcal{L}^*} = (C_i)^{(\mathcal{L}_{|C_i|})^*}$$
$$\subseteq (C_i)^{(\mathcal{L}_{k+2})^*} \subseteq P^{(\mathcal{L}_{k+2})^*} = P.$$

If there were an index i with $C_i = C_{i+1}$, i.e., C_i were $(\mathcal{L}_{|C_i|})^*$-closed, then C_i must be an intent or a pseudo-intent of $\mathbb{K}_{|C_i|}$. In particular, $Q^{I_{k+2}I_{k+2}} \subsetneq C_i \subseteq P$. If C_i were an intent, then also one of $\mathbb{K}_{k+2}$, which contradicts the maximality of Q. Hence C_i must be a pseudo-intent, and in particular one of $\mathbb{K}_{k+2}$ by Lemma 17. Due to the fact that Q is a maximal pseudo-intent below P, we may conclude that $C_i = P$. In summary, it follows that the sequence strictly converges to P (in finitely many steps if $P \setminus Q$ is finite), i.e., ends with P, and since each element is a candidate, P must occur as a candidate in $\mathbf{C}$.

Eventually, we have to consider the case where no pseudo-intent of $\mathbb{K}_{k+2}$ below P exists. In particular, then $\emptyset$ must be an intent of $\mathbb{K}_{k+2}$. As a consequence, $\emptyset$ is an intent of $\mathbb{K}_0$, too, as otherwise there would be an implication with premise $\emptyset$ in $\mathcal{L}$. Thus, the candidates $\{m\}$ where $m \in P$ have been inserted into $\mathbf{C}$. We may now define a sequence as above, but with $C_0 := \{m\}$ for an $m \in P$, and argue similarly as above. However, we have to additionally take care of the case $C_i = C_{i+1}$, as we may not use the maximality argument. Instead, assume that i is a minimal such index, and then merely continue the sequence with $C_{i+1} := C_i \cup \{m\}$ where $m \in P \setminus C_i$. This choice is suitable, since then C_{i+1} is a candidate, too, cf. Line 9. It follows that it is a sequence of candidates that ends with P, i.e., $P \in \mathbf{C}_{k+2}$. □

Theorem 18. *Let $\mathbb{D}$ be an (inaccessible) formal context with a finite attribute set, $\mathbb{K}$ be a finite subcontext of $\mathbb{D}$ such that $\mathbb{K}$ and $\mathbb{D}$ share the same attribute set, i.e., $\mathbb{K} \leq_M \mathbb{D}$, and χ an expert that is induced by $\mathbb{D}$ and answers questions in finite time. If Algorithm 1 is started on $(\mathbb{K}, \chi)$ as input, then it terminates, and returns a refinement $\mathbb{K}^\circ$ of $\mathbb{K}$ with $\mathsf{Int}(\mathbb{K}^\circ) = \mathsf{Int}(\mathbb{D})$ as well as a minimal implicational base of $\mathbb{D}$.*

Furthermore, there is no algorithm that computes a minimal implicational base of $\mathbb{D}$, but poses less questions to χ than Algorithm 1.

Proof. Termination is a consequence of finiteness of $\mathbb{K}$. If $\mathbb{K}$ is finite, then it has a finite attribute set M, and consequently there may only be finitely many candidates on each level. Furthermore, the computation of closures w.r.t. the operator $\mathcal{L}^*$ can always be obtained in finite time, since the implication set $\mathcal{L}$ consists of finitely many implications at any time during the algorithm's run. Obviously, also the intent closure $\cdot^{II}$ can be computed in finite time for finite contexts. Since each candidate is only used once to pose a question to the expert, it is not possible that the expert may return infinitely many counterexamples, and hence the adjusted context cannot grow to an infinite size.

The context $\mathbb{K}^\circ := \mathbb{K}_{|M|}$ of the final state is returned by the algorithm. It is readily verified that it contains the initial context $\mathbb{K}$ as a subcontext. Furthermore, due to the fact that $\mathbb{K}$ is itself a subcontext of $\mathbb{D}$, and during the algorithm's run only intents of $\mathbb{D}$ are added as new rows to $\mathbb{K}$, we conclude that $\mathsf{Int}(\mathbb{K}^\circ) \subseteq \mathsf{Int}(\mathbb{D})$.

By Proposition 16, it follows that in the final state $|M|$, $\mathcal{L}^\circ := \mathcal{L}_{|M|}$ is the canonical base of $\mathbb{K}^\circ$, i.e., $\mathsf{Imp}(\mathbb{K}^\circ) = \mathsf{Imp}(\mathcal{L}^\circ)$. Furthermore, $\mathsf{Imp}(\mathbb{D}) = \mathsf{Imp}(\chi)$ by Definition 2. Since all implications in $\mathcal{L}^\circ$ have been accepted by χ, we conclude $\mathcal{L}^\circ \subseteq \mathsf{Imp}(\chi)$, and hence $\mathsf{Imp}(\mathcal{L}^\circ) \subseteq \mathsf{Imp}(\chi)$. From $\mathsf{Int}(\mathbb{K}^\circ) \subseteq \mathsf{Int}(\mathbb{D})$ it follows that $\mathsf{Imp}(\mathbb{D}) \subseteq \mathsf{Imp}(\mathbb{K}^\circ)$. (If there were an implication that is valid in $\mathbb{D}$, but is not valid in $\mathbb{K}^\circ$, then a counterexample would exist which is an intent of $\mathbb{K}^\circ$, i.e., an intent of $\mathbb{D}$. Contradiction!) Consequently, the returned implication set $\mathcal{L}^\circ$ is indeed a minimal implicational base of $\mathbb{D}$. Since $\mathcal{L}^\circ$ is sound and complete for both $\mathbb{K}^\circ$ and $\mathbb{D}$, it follows that $\mathsf{Int}(\mathbb{K}^\circ) = \mathsf{Mod}(\mathcal{L}^\circ) = \mathsf{Int}(\mathbb{D})$.

The last claim is an immediate consequence of the fact, that $\mathcal{L}_{|M|}$ is a minimal implicational base. □

5   Discussion

Of course, it would be possible to utilize multiple experts in the default *Attribute Exploration* [4–7,13], but however this would not give any performance boost (if we assume that all experts answer immediately), as only one question in form of an implication is constructed at a time. If we compare Algorithm 1 with default *Attribute Exploration*, the order of the questions is different. They are enumerated in the lectic order, while the `ParallelAttributeExploration` enumerates w.r.t. increasing set cardinality of the premises. This means that on the one hand between two states of Algorithm 1 several implications can be processed in parallel, and on the other hand the difficulty of the questions (when measured in premise size) increases during the algorithm's run. In the default *Attribute Exploration* the difficulty of the questions varies during the algorithm's run, as they are constructed in the lectic order that does not respect set cardinality. Furthermore, the default algorithm cannot continue before the last posed question has been answered, but in contrast the parallel algorithm may process all posed questions with same premise cardinality in parallel. However, both algorithms return the same result for the same experts.

For an integration of several experts, there are the following options:

1. Randomly choose an (idle) expert, and pose the question to her.
2. Pose the question to all experts, and return the first answer.
3. Pose the question to all experts, and accept if all experts accept.
4. Pose the question to all experts, and accept if at least one expert accepts.

However, if we assume that all available experts are indeed induced by the formal context describing the domain of interest, i.e., have the same knowledge, then each of the four possibilities above would yield the same result, and hence it suffices to equally distribute the questions to all available experts. The four choices would only create different behaviours if the knowledge of the experts is not equivalent, or if the answering delays vary.

For instance, assume that there are experts $\chi_1, \ldots, \chi_n$ such that each χ_i is induced by a formal context $\mathbb{D}_i \leq_M \mathbb{D}$, and $\bigcup_{i=1}^{n} \mathbb{D}_i = \mathbb{D}$, i.e., each expert knows a part of the domain of interest, and no part of the domain of interest is unknown. Then an implication is valid in $\mathbb{D}$ if, and only if, it is valid in each $\mathbb{D}_i$. An induced expert χ of $\mathbb{D}$ is then obtained with the following definition: For an implication $X \to Y$, let $\chi \models X \to Y$ if $\chi_i \models X \to Y$ for all indices $i \in \{1, \ldots, n\}$, and otherwise define $\chi(X \to Y)$ as an arbitrary element of $\{C \mid \exists i \in \{1, \ldots, n\} : \chi_i(X \to Y) = C\}$, i.e., query all experts, and return the first counterexample, or accept otherwise.

6   Conclusion

We have considered the problem of *Parallel Attribute Exploration*, where a (minimal) implicational base for a domain of interest shall be computed in a parallel manner. The domain of interest is a formal context of which only some objects and their intents are known, and furthermore some experts are available that can correctly decide whether implications are valid. The introduced algorithm `ParallelAttributeExploration` is an extension of the algorithm

NextClosures [10,11], and a prototypical implementation is available [9]. It is planned to utilize it for a collaborative knowledge acquisition platform.

As a future step, the algorithm will be further extended to handle background knowledge, as this has been done by [4,13] for the default *Attribute Exploration* with lectic order. Furthermore, the algorithm will be generalized to the case where the data-set is described in terms of a closure operator in a (graded) complete lattice.

Acknowledgements. The author gratefully thanks the anonymous reviewers for their constructive hints and helpful remarks.

References

1. Angluin, D., Frazier, M., Pitt, L.: Learning conjunctions of horn clauses. Mach. Learn. **9**, 147–164 (1992)
2. Arias, M., Balcázar, J.L.: Canonical Horn representations and query learning. In: Gavaldà, R., Lugosi, G., Zeugmann, T., Zilles, S. (eds.) ALT 2009. LNCS, vol. 5809, pp. 156–170. Springer, Heidelberg (2009)
3. Borchmann, D.: Learning terminological knowledge with high confidence from erroneous data. Ph.D. thesis. Technische Universität Dresden (2014)
4. Ganter, B.: Attribute exploration with background knowledge. Theor. Comput. Sci. **217**(2), 215–233 (1999)
5. Ganter, B.: Two Basic Algorithms in Concept Analysis. FB4-Preprint 831. Darmstadt, Germany: Technische Hochschule Darmstadt (1984)
6. Ganter, B.: Two basic algorithms in concept analysis. In: Kwuida, L., Sertkaya, B. (eds.) ICFCA 2010. LNCS, vol. 5986, pp. 312–340. Springer, Heidelberg (2010)
7. Ganter, B., Wille, R.: Formal Concept Analysis: Mathematical Foundations. Springer, Heidelberg (1999)
8. Guigues, J.-L., Duquenne, V.: Famille minimale d'implications informatives résultant d'un tableau de données binaires. Mathématiques et Sci. Humaines **95**, 5–18 (1986)
9. Kriegel, F.: Concept Explorer FX. Software for Formal Concept Analysis (2010–2016). https://github.com/francesco-kriegel/conexp-fx
10. Kriegel, F.: NextClosures - Parallel Exploration of Constrained Closure Operators. LTCS-Report 15–01. Dresden, Germany: Chair of Automata Theory, Institute of Theoretical Computer Science, Technische Universität Dresden (2015)
11. Kriegel, F., Borchmann, D.: NextClosures: parallel computation of the canonical base. In: Proceedings of the 12th International Conference on Concept Lattices and Their Applications (CLA 2015), Clermont-Ferrand, France, 13–16 October 2015, pp. 181–192 (2015)
12. Maier, D.: The Theory of Relational Databases. Computer Science Press, Rockville (1983)
13. Stumme, G.: Attribute exploration with background implications and exceptions. In: Bock, H.-H., Polasek, W., et al. (eds.) Data Analysis and Information Systems. Studies in Classification, Data Analysis, and Knowledge Organization, pp. 457–469. Springer, Heidelberg (1996)

Graph-FCA in Practice

Sébastien Ferré[1]([⊠]) and Peggy Cellier[2]

[1] IRISA/Université de Rennes 1, Campus de Beaulieu, 35042 Rennes Cedex, France
`ferre@irisa.fr`
[2] IRISA/INSA Rennes, Campus de Beaulieu, 35042 Rennes Cedex, France
`cellier@irisa.fr`

Abstract. With the rise of the Semantic Web, more and more relational data are made available in the form of knowledge graphs (e.g., RDF, conceptual graphs). A challenge is to discover conceptual structures in those graphs, in the same way as Formal Concept Analysis (FCA) discovers conceptual structures in tables. Graph-FCA has been introduced in a previous work as an extension of FCA for such knowledge graphs. In this paper, algorithmic aspects and use cases are explored in order to study the feasibility and usefulness of G-FCA. We consider two use cases. The first one extracts linguistic structures from parse trees, comparing two graph models. The second one extracts workflow patterns from cooking recipes, highlighting the benefits of n-ary relationships and concepts.

Keywords: Formal Concept Analysis · Knowledge graph · Semantic Web · Graph pattern

1 Introduction

With the rise of the Semantic Web, more and more data are made available in the form of RDF graphs. More generally, *knowledge graphs* (e.g., RDF graphs, Conceptual Graphs) allow for the representation of complex information as a set of entities interlinked with binary, and possibly n-ary, relationships. A challenge is to discover conceptual structures in knowledge graphs, in the same way as Formal Concept Analysis (FCA) discovers conceptual structures in tables [6]. Relational Concept Analysis (RCA) [11], an extension of FCA, shows interesting results on relational data, in particular in software engineering [8].

Recently, Graph-FCA (G-FCA) [3] has been introduced as another extension of FCA for knowledge graphs. The specifity of G-FCA is to extract n-ary concepts from a knowledge graph using n-ary relationships. The intents of n-ary concepts are graph patterns with a focus on one or several nodes, i.e. *Projected Graph Patterns* (PGP). In practice, it allows to discover n-ary relational concepts. For instance, in a knowledge graph that represents family members with a "parent"

This research is supported by ANR project IDFRAud (ANR-14-CE28-0012-02).

O. Haemmerlé et al. (Eds.): ICCS 2016, LNAI 9717, pp. 107–121, 2016.
DOI: 10.1007/978-3-319-40985-6_9

relation, the "sibling" binary concept can be discovered, and described by a PGP as "a pair of persons having a common father and a common mother".

In this paper, algorithmic aspects of G-FCA and use cases are explored in order to study the feasibility and usefulness of G-FCA. We address the problem of an efficient generation strategy of concepts in order to avoid as much as possible duplication in computation and in the presentation of results. We have also conducted experiments on two use cases. The first one is the exploration of linguistic structures in parse trees, comparing two graph modellings. The second one is about the extraction of workflow patterns from cooking recipes

In the following, Sect. 2 recalls the main definitions of G-FCA. Section 3 discusses related work. Section 4 presents an algorithm to extract concepts from a graph context. Section 5 shows two use cases.

2   Graph-FCA

Graph Contexts. We here recall the main definitions and theoretical results of Graph-FCA (G-FCA) [3], and illustrate them with an example about the British royal family. Whereas FCA defines a formal context as an incidence relation between objects and attributes, G-FCA defines a *graph context* as an incidence relation between *tuples of* objects and attributes.

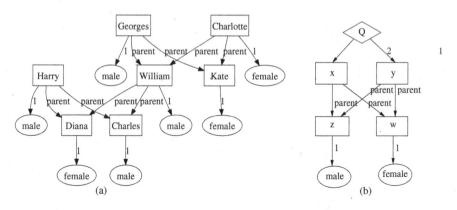

Fig. 1. (a) Graph context about the British royal family. Rectangles are objects, word-labelled links are binary edges, and ellipses are other edges. (b) PGP defining the "sibling" binary relation. Rectangles are variables, diamonds are projection tuples.

Definition 1 (graph context). *A graph context is a triple* $K = (O, A, I)$, *where O is a set of* objects, *A is a set of* attributes, *and $I \subseteq O^* \times A$ is an* incidence relation *between object tuples ($\bar{o} \in O^*$)[1] and attributes ($a \in A$).*

---

[1] Empty tuples are covered for the sake of generality but are not used in this paper.

The graphical representation of a graph context uses objects as nodes, incidence elements as hyper-edges, and attributes as hyper-edge labels. Note that attributes can be interpreted as n-ary predicates, and graph contexts as First Order Logic (FOL) models (without functions and constants). Different kinds of knowledge graphs, such as conceptual graphs, RDF graphs, and RCA contexts, can all be mapped easily to a graph context. Figure 1(a) shows the graphical representation of a small graph context about the British royal family. The objects are the royal family members (e.g., Harry, Georges). They are represented as rectangles. The attributes are a binary relation "parent" and two unary relations "male" and "female". The edges ((Harry,Charles),parent), ((Georges,Kate),parent), and ((Harry),male) belong to the incidence relation. More generally, a binary edge $((x, y), a)$ is represented by an edge from x to y labelled by a. Other edges $((x_1, \ldots, x_n), a)$ are represented by ellipses labelled by a, and having an edge labelled i to each node x_i.

Graph Patterns and Projections. *Graph patterns* generalize the incidence relation of a graph context by taking its nodes in an infinite set of variables $x, y, \ldots \in \mathcal{V}$. Therefore, a graph context can be seen as a graph pattern by abstracting its objects as variables. A *Projected Graph Patterns* (PGP) is a graph pattern with a tuple of distinguished variables.

Definition 2 (graph pattern and PGP). *A graph pattern $P \subseteq \mathcal{V}^* \times A$ is a set of directed hyper-edges with variables as nodes, and attributes as labels.*

A projected graph pattern (PGP) is a couple $Q = (\overline{x}, P)$ where P is a graph pattern, and $\overline{x} \in \mathcal{V}^*$ is called the projection tuple. *The* arity *of a PGP is the length of $\overline{x}$. We note $\mathcal{Q}_k$ the set of PGPs having arity k.*

A key aspect of G-FCA is that closure does not apply directly to graph patterns but to PGPs. PGPs are analogous to anonymous definitions of FOL predicates, and to SPARQL queries. They play the same role as sets of attributes in FCA, i.e. as concept intents. Figure 1(b) shows a PGP defining the "sibling" binary relation as two persons sharing a male parent (father) and a female parent (mother). The projection tuple is (x, y). Its graphical representation is a diamond node pointing to each projected variable.

Set operations are extended from sets of attributes to PGPs. PGP inclusion $\subseteq_q$ is based on graph homomorphisms [7]. It is similar to the notion of *subsumption* on queries [2] or rules [10]. PGP intersection $\cap_q$ is defined as a form of graph alignment, where each pair of variables from the two patterns becomes a variable of the intersection pattern. It corresponds to the *categorical product* of graphs (see [7], p. 116).

Definition 3 (PGP inclusion). *A k-PGP $Q_1 = (\overline{x_1}, P_1)$ is included in a k-PGP $Q_2 = (\overline{x_2}, P_2)$, denoted by $Q_1 \subseteq_q Q_2$, iff there exists a mapping ϕ from P_1-nodes to P_2-nodes s.t. $\overline{\phi(x_1)} = \overline{x_2}$, and for every edge $(\overline{y}, a) \in P_1$, $(\overline{\phi(y)}, a) \in P_2$. Therefore, ϕ is an homomorphism from P_1 to P_2 that preserves the projection tuple. When $Q_1 \subseteq_q Q_2$ and $Q_2 \subseteq_q Q_1$, they are said equivalent ($Q_1 \equiv_q Q_2$).*

Definition 4 (PGP intersection). *Let ψ be an injective mapping from pairs of variables to variables. The intersection two k-PGPs $Q_1 = (\overline{x_1}, P_1)$ and $Q_2 = (\overline{x_2}, P_2)$, denoted by $Q_1 \cap_q Q_2$, is defined as $Q = (\overline{x}, P)$, where $\overline{x} = \psi(x_1, x_2)$, and $P = \{(\psi(y_1, y_2), a) \mid a \in A, (\overline{y_1}, a) \in P_1, (\overline{y_2}, a) \in P_2, |\overline{y_1}| = |\overline{y_2}|\}$.*

Object Relations. FCA sets of objects are extended to *object relations*, i.e. sets of tuples of objects $R \subseteq O^k$, for some arity k. We note $\mathcal{R}_k$ the set of relations with arity k. For instance, $\{(Charles, William), (Charles, Harry), (William, Georges)\}$ is an object relation with arity 2. Object relations are analogous to query answers in SPARQL. Object relations form a powerset lattice for each arity.

Galois Connection and Graph Concepts. Based on previous definitions, the following Galois connection can be defined and proved between PGPs and object relations (see [3] for the proof). The connection from PGP to object relation is analogous to query evaluation, and the connection from object relations to PGP to relational learning [10]. In the definitions of Q' and R' below, the PGP $(\overline{o}, I) \in \mathcal{Q}$ represents the description of an object tuple $\overline{o}$ by the relative position of objects in the whole incidence relation I.

Theorem 1 (Galois connection). *Let $K = (O, A, I)$ be a graph context. For every arity k, the following pair of mappings between PGPs $Q \in \mathcal{Q}_k$ and object relations $R \in \mathcal{R}_k$ forms a Galois connection.*

$$Q' := \{\overline{o} \in O^k \mid Q \subseteq_q (\overline{o}, I)\} \qquad R' := \cap_q \{(\overline{o}, I)\}_{\overline{o} \in R}$$

From there, concepts can be defined in the usual way, and proved to be organized into lattices. The only restriction compared to FCA is that the concept lattices may not be complete but this has no practical impact when data is finite.

Definition 5 (graph concept). *A graph concept with arity k is a pair (R, Q), made of an object relation $R \in \mathcal{R}_k$ (the extent) and a PGP $Q \in \mathcal{Q}_k$ (the intent), such that $R = Q'$ and $Q \equiv_q R'$.*

Figure 2 displays a compact representation of the graph concepts about the British royal family. Each node x identifies a unary concept (e.g., Q3e) along with its extent (here, $\{Charlotte, Georges, Harry, William\}$). The concept intent is the PGP $((x), P)$, where P is the subgraph containing node x and all white nodes (called the *pattern core*, i.e. the nodes that appear in all represented concepts). Concept Q3e is concept "child", which, in the graph context, always has a known father and mother, which always have a son. Note that the son maybe either the child's brother or the child himself because homomorphisms need not be injective. Concept Q1a is concept "female person". Concept Q4i uniquely characterizes Charlotte in the graph context as being female, having parents, paternal grand-parents, and a brother. Note that there is no concept that uniquely characterizes Harry because his description is a subset of William's description; hence concept Q4g gathering William and Harry. In total, there

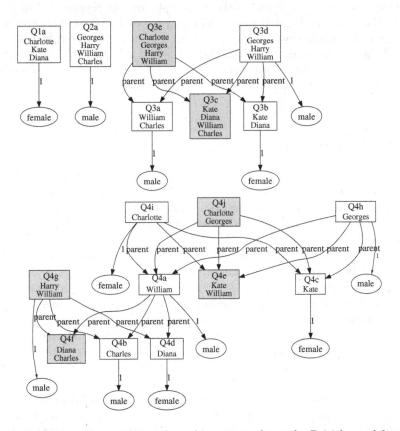

Fig. 2. Compact representation of graph concepts about the British royal family.

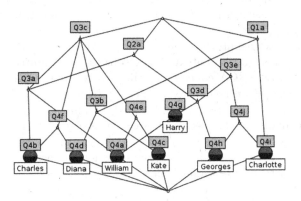

Fig. 3. Lattice of the unary concepts about the British royal family.

are 19 unary concepts (top and bottom concepts are not represented in Fig. 2). Figure 3 shows the lattice structure of all unary concepts. Binary concept intents are obtained by picking two nodes to form a projection tuple, and by taking the union of the two patterns. For example, concept (Q3a,Q3b) is concept "couple sharing a son". Concept (Q3e,Q3d) defines the relationship "having a brother", if we exclude self-relationships. Ternary and other n-ary concepts are formed likewise.

3   Related Work

G-FCA graph patterns bear much similarity with Conceptual Graphs (CG) [1, 13], and we re-used their graphical notation. We adopted a slightly simpler formalization by not distinguishing between concept types, relation types, and individual markers, which are all modeled with attributes in G-FCA for uniformity. The semantics of knowledge graphs (e.g., CG type hierarchies, RDF Schema) is not natively handled but FCA techniques like scaling can easily be applied to G-FCA. Those differences are minor, and the novelty of G-FCA lies in projected graph patterns (PGPs), PGP intersection, and concept formation from a knowledge graph. For comparison, reasoning with CGs is mostly based on graph homomorphisms, typically between a source graph and a query graph.

G-FCA concept formation works differently from graph mining approaches [9, 14,15] because they generally consist in finding frequent substructures in a collection of graphs (e.g., molecules), and they use subgraph isomorphism instead of homomorphisms.

Previous FCA extensions, Logical Concept Analysis (LCA) [4] and Pattern Structures (PS) [5], have definitions for the Galois connection that look much like those of G-FCA (Theorem 1). However, in those extensions, descriptions only apply to single objects, and are independent one from the other. In G-FCA, the whole knowledge graph serves as a description, not only for single objects but also for tuples of objects. This allows for describing n-ary relationships between objects, as well as discovering new relationships (concepts) as complex combinations of primitive ones.

Another FCA extension, Relational Concept Analysis (RCA) [11], also discovers concepts in a graph context. RCA contexts are limited to unary and binary attributes, and RCA concepts are limited to unary concepts. However, the main difference lies in the nature of concept intents: (possibly infinite) rooted tree patterns instead of projected graph patterns. This implies that interesting cycles in data cannot be expressed in RCA concepts (e.g., concept Q3e in Fig. 2). Other advantages of G-FCA are (1) a declarative, rather than iterative, characterization of concepts, and (2) a self-contained graph-based representation of concept intents instead of cascading references to concepts.

4   Computation and Presentation of Graph Concepts

The computation of graph concepts is challenging because of the complexity of computing with graphs. The fact that PGP inclusion is based on graph homomorphism rather than on subgraph isomorphism like in most other work on graph patterns [9,15] is both an advantage and a drawback. The advantage is that every intersection of two PGPs is a PGP, so that it is not necessary to reason on sets of PGPs for computing concept intents. The drawback is that an intersection $Q_1 \cap_q Q_2$ may be larger than both Q_1 and Q_2 when an object has several edges with the same attribute: e.g., a parent of several children. Another difficulty is that it is more difficult to get canonical representations of PGPs compared to FCA sets of attributes. Indeed, two PGPs may be equivalent although their graph patterns are not isomorphic: e.g., graphs **H**, **G1**, and **G2** in Fig. 4. In the following, we first describe the naive version of bottom-up generation of concepts up to some arity k (Sect. 4.1). We then sketch an algorithm to factorize computations, and to generate a *concept basis* as a subset of unary concepts from which other concepts can be derived by simple operations (Sect. 4.2).

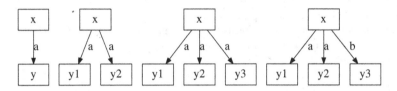

Fig. 4. Graphs **H** and **G1** are retracts of graph **G2** but not **G3**. Graph **H** is the core of graphs **G1** and **G2**. Hence, graphs **H**, **G1**, and **G2** are equivalent.

4.1   Naive Generation of Concepts

The adopted strategy is to generate concept intents in a bottom-up fashion, based on the second half of the Galois connection: $R' := \cap_q \{(\bar{o}, I)\}_{\bar{o} \in R}$. The principle for each arity k is to start from the set of all descriptions of k-tuples of objects $\{(\bar{o}, I) \mid \bar{o} \in O^k\}$, and to close it by application of PGP intersection $\cap_q$. It is easy to see why this naive generation is far from optimal. For example, in the royal family context, the generation of the unary concepts from the 7 objects already generates $C_7^2 = 21$ PGP intersections, and hence 21 alignments of the incidence relation on itself. It gets exponentially worse with concept arity increasing.

In order to detect when an intent has already been generated, each PGP $Q = (\bar{x}, P)$ (including object tuple descriptions) must be given the canonical representation of its equivalence class. That canonical representation is computed in two steps. First, the minimal *retract* R of the graph pattern P that contains the projected nodes $\bar{x}$ must be found. Second, the nodes of R are numbered in a canonical way assuming a fixed ordering of attributes. Roughly, a *retract* of a graph G is a subgraph of G that conveys the same information (for details

see [7], p. 112). In Fig. 4, graphs **H** and **G1** are retracts of graph **G2** but not
G3. Indeed, stating several times that x is in a a-relation to something adds
nothing to stating it once because variables $y1$, $y2$, and $y3$ can map to the same
object in the graph context. On the contrary, **G3** states that x is both in a a
relation and a b-relation, and cannot retract to **H**: edge $((x, y3), b)$ cannot fold
onto edge $((x, y2), a)$. A *core* is a minimal retract. In Fig. 4, graph **H** is the core
of graphs **G1** and **G2**. If there are several cores, any of them can be taken as
they are isomorphic.

4.2   Efficient Generation of a Concept Basis

In this section, we sketch an algorithm for a more efficient generation of concepts
(Algorithm 1). The objective is not quantitative performance (this is let to future
work), but *qualitative* performance. By that we mean the orderly generation of
concepts avoiding as much as possible duplication both in computation and in
presentation of results.

Algorithm 1. Generation of concepts

Require: $K = (O, A, I)$ is a graph context
Ensure: *Concepts* is the concept basis, a set of unary concepts $(R, Q)@P$ where each
  concept intent is presented as a subgraph of pattern P
1: $Concepts \leftarrow \emptyset$
2: $Patterns \leftarrow \{I\}$ // a queue of patterns to process
3: **for all** $P \in Patterns$ **do**
4:    **for all** new $P_a \in ConnectedComponents(P \cap_q I)$ **do**
5:      $P_a \leftarrow removeDuplicateNodes(P_a)$
6:      **for all** new $P_b \in Retracts(P_a)$ **do**
7:        $X \leftarrow P_b.nodes \setminus \bigcup_{R \in Retracts(P_a)|R \subsetneq P_b} R.nodes$ // nodes inducing P_b
8:        **if** $X \neq \emptyset$ **then**
9:          **for all** $x \in X$ **do**
10:            $Q \leftarrow ((x), P_b); R \leftarrow Q'$ // intent and extent
11:            $Concepts \leftarrow \{(R, Q)@P_a\} \cup Concepts$
12:          **end for**
13:          $Patterns \leftarrow \{P_b\} \cup Patterns$ // queuing P_b for intersection
14:        **end if**
15:      **end for**
16:    **end for**
17: **end for**

Generation of Graph Patterns. Instead of generating PGPs directly by
PGP intersection, we first generate alignments (categorical products) of graph
patterns ($P_1 \times P_2$), ignoring at this stage projection tuples (Step 4). Thus, the
incidence relation is aligned onto itself only once ($I \times I$). The graph product
may have several connected components. However, if a concept has an intent
whose pattern is made of several connected components, then each component
is closed, hence forms a concept intent. Therefore, non-connected concepts could

be composed from connected ones, and we choose to only generate connected PGPs, and hence only connected patterns. To summarize, whenever a product of two connected patterns is computed, each connected component $P_a^{\cdot}$ of the product becomes an input for the next stage (Step 4).

Generation of Concept Intents. For each connected graph pattern P_a, and for each projection tuple $\overline{x}$ taken from P_a-nodes, the PGP $(\overline{x}, P_a)$ is a concept intent. The number of projection tuples can be reduced by abstracting over the tuple ordering, and by considering projection sets instead of projection tuples. Indeed, every permutation of the projection tuple of a concept intent obviously leads to another concept intent. For each projection set X, the PGP (X, P_a) can be minimally retracted to (X, P_b) by choosing the smallest retract P_b of P_a that contains nodes X. We say that X *induces* retract P_b. The generation of PGPs can be optimized for two reasons. First, several projection sets may induce the same retract; second, if a projection set X induces retract P_b, then every projection set Y of P_b-nodes that includes X induces the same retract P_b. A more efficient strategy is therefore to first generate all retracts of P_a (Step 6), and then for each retract P_b to find the minimal projection sets that induce P_b (Step 7, after second optimization below). A non-minimal projection set could simply be obtained by extending it with other nodes of P_b. For example, from unary concept Q3e in Fig. 2, one could define the binary relationship "brother" by extending the projection set to Q3d. Note that all retracts P_b contains the core P_c of P_a as a subgraph. Generating all retracts amounts to computing all matches of P_a onto itself. Finding the minimal projection sets amounts to enumerating the minimal subsets of P_b-nodes that are not included in any retract that is smaller than P_b. In this way, all concept intents sharing a same graph pattern are found together (Steps 10–11). For an even more compact presentation, concepts can further be grouped by the core P_c of their pattern. Finally, each new retract P_b is fed as input to the previous stage (Step 13).

Optimization in Case of Symmetries. In case of 1-n or n-n relationships (e.g., persons having several children), the product graph patterns P_a often exhibit duplications in the sense that several non-adjacent nodes play exactly the same role in the pattern: e.g., nodes $y1$ and $y2$ in graph **G3** of Fig. 4. Duplications can lead to a combinatorial explosion in the generation of all retracts of a pattern P_a. The optimization consists in first partitioning P_a-nodes by grouping those that play the same role, and then keeping only one node out of each group (Step 5). The only consequence is to miss the concept intents that have several projected nodes from the same group. However, those intents could easily be retrieved by duplicating projected nodes in generated intents. For example, the binary concept "parent couple" (Q3c,Q3c) is obtained by duplicating the unary concept "parent" Q3c (see Fig. 2).

Optimization by Reduction to Unary Concept Intents. It can be proved that for a retracted pattern P_b induced by a projection set X, P_b is the union of the retracts induced by each projected node $x \in X$. Conversely, the union of the retracts induced by a row of projected nodes $x \in X$ is the retract induced by X. Therefore, n-ary intents could be derived from the set of unary intents extracted from a connected pattern P_a (Step 9). For that reason, it is important, in the representation of generated intents, to collectively show the patterns of unary intents as a subgraph of P_a (hence the notation $@P_a$ at Step 11), rather than individually.

4.3 Implementation

We have implemented the above algorithm as a prototype in about 1700 lines of OCaml[2]. It takes as input a graph context, i.e. a set of directed hyper-edges. It returns as output a compact representation of the concept basis, like in Fig. 2. For each pattern, its core nodes are shown in white while other projection nodes are shown in grey. Each node represents a unary concept, and is identified by the pattern number and a letter (e.g., **Q1a**). The graph pattern of the concept intent is the subgraph induced by the set of nodes made of: the projected node x, the core nodes, and possibly other nodes between x and core nodes (indicated between brackets after x's label). N-ary concepts can be derived by picking several nodes, and merging their patterns. The prototype has options for specifying maximum intent pattern size (nb. of nodes), for computing and displaying unary concept extents, and for formatting results in graphical form (.dot file) or in Prolog-like textual form.

5 Use Cases

In this section, two use cases are presented in order to show the interest of G-FCA concepts and their PGPs. The first one extracts linguistic structures from parse trees, comparing two graph modellings. The second one extracts workflow patterns from cooking recipes, highlighting the benefits of n-ary relationships.

5.1 Extraction of Concepts in Parse Trees

We have conducted experiments on the poem of the french poet Charles Baudelaire named *"La beauté"*[3]. The goal of those experiments is the extraction of syntactic structures appearing in the text. We use the french chunker of the tool Treetagger [12][4] in order to automatically build the parse tree of the text.

From the computed parse tree, we propose two modellings to represent the poem: the first model represents the parse tree without taking into account the order between words, only the composition relation, whereas the second model explicitly encodes the order between words in addition to the composition relation. Both modellings are presented in the sequel and then the extracted patterns are discussed.

---

[2] Source code, datasets, and concept sets at https://bitbucket.org/sebferre/g-fca.

[3] In "Les Fleurs du mal". Charles Baudelaire. 1857.

[4] http://www.cis.uni-muenchen.de/~schmid/tools/TreeTagger/.

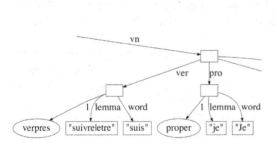

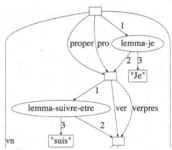

Fig. 5. Excerpt of the composition modelling for the phrase "Je suis".

Fig. 6. Excerpt of the sequential modelling for the phrase "Je suis".

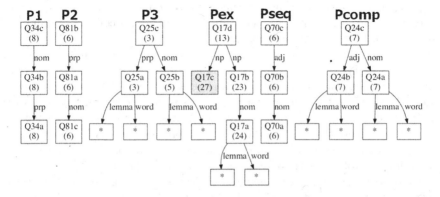

Fig. 7. Concepts extracted from the poem.

Composition Modelling. In the first representation of the text, edges represent a "contain" relation between phrases and words. For instance, Fig. 5 shows an excerpt of the graph of the sentence *"Je suis belle, ô mortels !"*. In this representation, the sentence ("s") is described as containing a verbal nucleus ("vn") which in turn contains: (1) a verb ("ver") which has a word ("suis"), a lemma ("suivre" or "être") and another part-of-speech (POS) information ("verpres", i.e. verb in present); (2) a pronoun ("pro") which has a word ("je"), a lemma ("je") and another part-of-speech information ("proper", i.e. personal pronoun); and two other words (a punctuation and an adjective). Note that the part-of-speech tags ("ver", "vn", "proper", etc.) are Treetagger tags.

Sequential Modelling. In the second representation, nodes represent positions between words, and edges represent phrases and words between those positions. POS tags and lemmas are used as attributes. For instance, Fig. 6 shows an excerpt of the graph of the sentence *"Je suis belle, ô mortels !"*. For instance, the word "Je" is represented by three edges from the top node: attributes "proper", "pro" (POS information), and "lemma-je" (lemma). The fact that edge "vn" overlaps edges "pro" and "ver" represents composition.

Discussion About Extracted Patterns. When extracting concepts from both representations, we note that more patterns are extracted from the sequential model than from the composition model. For instance, with parameter $maxsize = 10$ (maximum number of nodes per PGP), 284 patterns are extracted in the sequential model instead of 68 patterns for the composition model[5]. Indeed, in the sequential model, the graph structure is rigid, the order between words is really important. When concepts are extracted, the sequential model thus generates more distinct patterns. For instance, let us consider P1, P2 and P3 in Fig. 7. P1 and P2 are extracted from the sequential model whereas P3 is extracted from the composition model. The three patterns represent the association between a preposition and a noun. However the composition modelling generates only one concept when the sequential modelling generates two concepts taking into account the ordering of the two words.

Some structural information about the text can be retrieved in the concepts. For instance, in Fig. 7, P_{ex}, extracted from the composition model, exhibits an obvious pattern in the poem, i.e. a noun phrase (np) which contains a noun (*nom*). The size of the extent of unary concepts is given between brackets. Concept (Q17b) can be read as "a noun phrase that contains a noun and that belongs to something". Note that the size of the extension of (Q17b) is 23 objects. Concept (Q17a) can be read as "a noun that belongs to a noun phrase that belongs to something". Note that the size of the extension of (Q17a) is 24 objects. The size of (Q17a) is thus greater than the size of (Q17b), it means that an object in the extension of Q17b contains not only one but two nouns ("un rêve de pierre"). Concept (Q17c) can be read as "a noun phrase that belongs to something that contains a noun phrase that contains a noun". Note that the size of the extension of (Q17c) is 27 objects. This concept is interesting, indeed it exhibits the fact that a noun phrase does not necessarily contain a noun in this text, for instance it can also be a pronoun: "*où*" (where), "*chacun*" (everyone). It also shows that two noun phrases can be found in the same structure.

The information conveyed by the two modellings are not the same, however both representations are interesting. For instance, let us consider the two patterns P_{seq} and P_{comp} in Fig. 7 that represent the association of a noun and an adjective. The size of the extension of concepts in P_{seq} is 6 and the size of the extension of concepts in P_{comp} is 7. Indeed, the six phrases that match P_{seq} also match P_{comp}. However the phrase "*toutes choses plus belles*", which contains an adverb ("*plus*") between the noun and the adjective, only match Q_{comp}. The sequential modelling allows to take into account the order between words, it is more accurate whereas the composition modelling allows for more general patterns. The choice of the more appropriate modelling depends on the task.

---

[5] Note that, the extraction of the 284 patterns in the sequential model takes about 4 s and the extraction of the 68 patterns in the composition model takes about 20 s.

5.2   Extraction of Concepts in Recipes

We have also conducted experiments on recipes. Four recipes are modelised: chocolate apple pie, strawberry-apple pie, mango-coconut pie and condoeuvre (Rhubarb pie or gooseberry pie). In this example, n-ary relations are used to represent temporal constraints between actions, and entities manipulated by actions. For instance, "put_on" is a quaternary relationship relating (1) start, (2) end, (3) object (e.g., "fruit"), and (4) destination (e.g., "pastry"). All action attributes use a similar schema (e.g., "cut", "bake_for"); other attributes represent types of ingredients or ustensils (e.g., "cream", "dish"). From those four recipes, 43 patterns are extracted in less than 1s. An excerpt is given in Fig. 8. Some patterns are very small and very frequent as (a) in Fig. 8. They represent ingredients (e.g., sugar, cream) or atomic actions (e.g., "put something on something", "pour on"). Some patterns are larger but less frequent as (b) in Fig. 8. They represent refinements of previous patterns or very specific actions (e.g., "pour cream on something"). Finally, some patterns are large and still frequent as (c) in Fig. 8. They correspond to the abstraction of many recipes. In this example, the pattern represents an abstraction of a pie recipe. It means: "cut the fruit in order to put it on something (often a pastry), which is put on a dish, which is baked, after preheating the oven, in order to obtain a pie". As an example of n-ary concept, the ternary concept (Q13j,Q13a,Q13c) can be used to relate a pie (Q13j) to the kind of base (Q13a) and fruit (Q13c) it is made of, while abstracting over other details of the recipe.

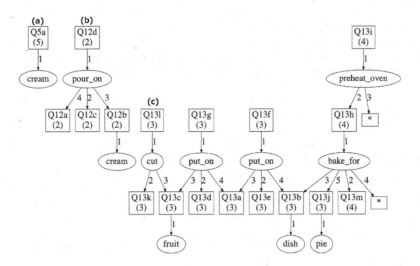

Fig. 8. Three concepts extracted from the recipes.

6 Conclusion

In this paper, we have proposed an algorithm to compute graph concepts in knowledge graphs. In particular, we tackle the problem of the generation and representation of the PGPs representing concept intents in a compact way. We also describe two use cases. The first use case, on textual data, allows to discuss two kinds of modelling (with or without sequentiality). The other use case, on cooking recipes, shows the interest of G-FCA for n-ary relations. With those two use cases we have seen that PGPs offer expressive patterns that can mix sequentiality, temporality, and composition thanks to n-ary relations. However, the set of extracted concepts can be large. Further work is to find a way to facilitate, for a user, navigation among them.

References

1. Chein, M., Mugnier, M.L.: Graph-Based Knowledge Representation: Computational Foundations of Conceptual Graphs. Advanced Information and Knowledge Processing. Springer, London (2008)
2. Chekol, M.W., Euzenat, J., Genevès, P., Layaïda, N.: SPARQL query containment under RDFS entailment regime. In: Gramlich, B., Miller, D., Sattler, U. (eds.) IJCAR 2012. LNCS, vol. 7364, pp. 134–148. Springer, Heidelberg (2012)
3. Ferré, S.: A proposal for extending formal concept analysis to knowledge graphs. In: Baixeries, J., Sacarea, C., Ojeda-Aciego, M. (eds.) ICFCA 2015. LNCS, vol. 9113, pp. 271–286. Springer, Heidelberg (2015)
4. Ferré, S., Ridoux, O.: A logical generalization of formal concept analysis. In: Ganter, B., Mineau, G.W. (eds.) ICCS 2000. LNCS, vol. 1867, pp. 371–384. Springer, Heidelberg (2000)
5. Ganter, B., Kuznetsov, S.O.: Pattern structures and their projections. In: Delugach, H.S., Stumme, G. (eds.) ICCS 2001. LNCS (LNAI), vol. 2120, pp. 129–142. Springer, Heidelberg (2001)
6. Ganter, B., Wille, R.: Formal Concept Analysis: Mathematical Foundations. Springer, Heidelberg (1999)
7. Hahn, G., Tardif, C.: Graph homomorphisms: structure and symmetry. In: Hahn, G., Sabidussi, G. (eds.) Graph Symmetry, pp. 107–166. Springer, Dordrecht (1997)
8. Huchard, M.: Analyzing inheritance hierarchies through formal concept analysis: a 22-years walk in a landscape of conceptual structures. In: MechAnisms on SPEcialization, Generalization and inHerItance (MASPEGHI), pp. 8–13. ACM (2007)
9. Kuznetsov, S.O., Samokhin, M.V.: Learning closed sets of labeled graphs for chemical applications. In: Kramer, S., Pfahringer, B. (eds.) ILP 2005. LNCS (LNAI), vol. 3625, pp. 190–208. Springer, Heidelberg (2005)
10. Muggleton, S., Raedt, L.D.: Inductive logic programming: theory and methods. J. Logic Program. **19**(20), 629–679 (1994)
11. Rouane-Hacene, M., Huchard, M., Napoli, A., Valtchev, P.: Relational concept analysis: mining concept lattices from multi-relational data. Ann. Math. Artif. Intell. **67**(1), 81–108 (2013)
12. Schmid, H.: Probabilistic part-of-speech tagging using decision trees. In: International Conference on New Methods in Language Processing (1994)

13. Sowa, J.: Conceptual Structures: Information Processing in Man and Machine. Addison-Wesley, Reading (1984)
14. Washio, T., Motoda, H.: State of the art of graph-based data mining. SIGKDD Explor. Newsl. **5**(1), 59–68 (2003). http://doi.acm.org/10.1145/959242.959249
15. Yan, X., Han, J.: Closegraph: mining closed frequent graph patterns. In: ACM International Conference on Knowledge Discovery and Data Mining (SIGKDD), pp. 286–295. ACM (2003)

A Semiotic-Conceptual Analysis
of Conceptual Learning

Uta Priss[✉]

Zentrum für erfolgreiches Lehren und Lernen,
Ostfalia University of Applied Sciences, Wolfenbüttel, Germany
http://www.upriss.org.uk

Abstract. While learning mathematics or computer science, beginning
students often encounter significant problems with abstract concepts. In
both subjects there tend to be large numbers of students failing the class
or dropping out during the first semesters. There is a substantial existing
body of literature on this topic from a didactic perspective, but in our
opinion an investigation from a semiotic-conceptual perspective could
provide further insights and specifically analyse the difficulties encountered when learning abstract concepts. This means that both the com-
plexities of the representations of abstract concepts and the conceptual
content itself are modelled and investigated separately and in combination with each other. In our opinion a semiotic analysis of the representations is often missing from didactic theories. And in particular, as
far as we know, there are not yet any formal mathematical approaches
to modelling learning difficulties with respect to semiotic and conceptual structures. Semiotic-Conceptual Analysis (SCA) as presented in this
paper aims to fill that niche.

1 Introduction

Semiotic-Conceptual Analysis (SCA) was inspired by Charles S. Peirce's triadic
definition of signs but does not claim to present an exact formalisation of his
ideas. A more detailed discussion of how SCA relates to Peirce was provided
by Priss (2015) and shall not be repeated in this paper. Peirce's semiotics was
aimed at analysing signs occurring in natural communication where representamens (physical representations of signs) are visible or audible (in form of words,
gestures and so on) but denotations (meanings) and mental interpretations can
only be speculated about. Nowadays computer programs are examples of sign
communication where every aspect of the signs, their representations, inputs,
outputs, states and runtime behaviour can be documented in minute detail.
Furthermore modern programming languages display a variety of complex structures (such as abstract data types, object orientation or functions as first class
objects) which are probably far beyond the complexity that is expressible within
natural languages. Thus formal languages are an interesting domain for semiotic
analyses. Analysing program source code was one of the motivations for developing SCA (Priss 2015). Another interesting semiotic aspect is how people interact

O. Haemmerlé et al. (Eds.): ICCS 2016, LNAI 9717, pp. 122–136, 2016.
DOI: 10.1007/978-3-319-40985-6_10

with such formal representations and, also, what difficulties students encounter when they are learning to interpret and use formalisms in mathematics and computer science. That is the focus of this paper. It shows how SCA can serve as a tool for exploring and highlighting difficulties within representations and their underlying abstract content. While there are already many existing approaches to semiotics, our goal is to develop a theory that builds on mathematical formalisations of signs and of concepts in the sense of Formal Concept Analysis (FCA[1]). To our knowledge such an FCA-based theory of semiotics does not yet exist elsewhere.

SCA defines signs as instances of a triadic relation consisting of representamens, denotations (or meanings) and interpretations. Interpretations are functions that map representamens into denotations. Peirce uses the term 'object' instead of 'denotation' and 'interpretant' instead of 'interpretation'. Priss (2015) explains why SCA adopts a different terminology. It should be emphasised that similar to how FCA uses its notions, the terms 'representamen', 'denotation' and 'interpretation' indicate structural positions within the formal model presented by SCA. As will be explained below such notions are 'anonymous signs' in the terminology of SCA. When explaining FCA to non-mathematicians one sometimes encounters criticisms such as 'what you are calling concepts are not concepts'. But from a mathematical viewpoint, 'concept' is just a name for a structure. It does not mean anything other than what is defined. Only in applications such notions acquire additional meaning which can be investigated with respect to their appropriateness in other domains. For example, whether or not SCA provides a 'semiotic analysis' in the ordinary sense depends on how it is used in an application. From a teaching perspective, the use of anonymous signs may be one of the core difficulties that students encounter when they learn mathematics. Students often associate concrete meanings with anonymous signs or fixate on specific representations instead of realising that the meaning of the sign in question is purely structural.

Our goal for SCA is to describe a semiotic theory that is applicable to all signs and all kinds of representamens. Each of the three components of signs (representamens, denotations and interpretations) has conceptual structures and some form of similarity. For SCA we use concept lattices but the core approach and terminology of SCA would still be applicable even if, for example, conceptual graphs were used instead of concept lattices. One of the core questions is whether similar representamens have similar denotations under similar interpretations. From an educational viewpoint one can investigate, for example, whether the interpretations used by a student are similar to the interpretations used by a teacher. If a student has 'understood' a concept then he or she should use the signs relating to this concept in a similar manner as a teacher.

---

[1] Because Formal Concept Analysis (FCA) has been presented many times at this conference, this paper does not provide an introduction to FCA but there is an example with some explanation in Sect. 5. Further information about FCA can be found, for example, on-line (http://www.fcahome.org.uk) and in the main FCA textbook by Ganter and Wille (1999).

The next section explains briefly why it is useful to define different notions of similarity for signs (such as synonymy) instead of just defining equality. Section 3 of this paper provides a brief overview over other existing theories that are relevant in this context. Section 4 establishes the formal definitions of SCA. Sections 5 and 6 demonstrate how SCA can be used for didactic applications. The paper finishes with a concluding section.

2   Equality and Similarity of Signs

One of the challenges for semiotics is that it is common to ignore certain aspects of signs in everyday language where a sign and its representamen are not always clearly distinguished. In mathematics, representamens tend to be ignored and equality tends to be denotational. For example, $x = 5$ means that the value of x is 5. Obviously the representamens, a letter x and a number 5, are different. This raises the question as to what it means for two signs to be equal to each other. It should be mentioned that a sign can be observed at different levels of granularity corresponding to a sentence, word or character. For example if someone expresses the sentence "I just bought an apple" on two consecutive days then it is a question whether this constitutes a single sign or two different signs. Most likely two different apples will be involved. Nevertheless, as elaborated by Priss (1998) the meaning of a word is not an object (a real apple) but a concept ('apple'). This concept could still be the same on both days. Because signs are triadic, however, two signs can only be equal if their representamens, denotations and interpretations are all equal (or at least equivalent). If the interpretations[2] contain information about the spatial and temporal context in which the sign was used, then the two sentences about eating an apple must be different signs. If the interpretations are less detailed, then it could be the same sign used on different days. We use the term 'equinymy' to describe this case where signs have equivalent representamens and equal denotations but possibly different interpretations. How interpretations are chosen is up to the person who uses SCA for modelling. Thus whether or not the sentence above corresponds to one sign or to two equinymous signs is a consequence of modelling decisions. If on the first day 'apple' refers to fruit and on the second day to a computer, then two different interpretations (and thus two signs) are required because interpretations are functions. In that case the term 'homograph' describes signs that have the same representamen but totally different meanings.

Synonymous signs have different representamens but similar meaning. Table 1 shows four pairs of representamens with similar meanings. Any statements that can be made about these always depend on interpretations. The first two representamens are distinguished by their font. SCA uses tolerance or equivalence

---

[2] 'Interpretation' in SCA is an anonymous sign and refers to a function. 'Having a different interpretation' in SCA means using a different function. Different interpretations can still lead to the same denotation. This is different to how 'interpretation' is used in ordinary language.

relations[3] which express which representamens are considered to be similar or equivalent. If the representamens in the first row are considered equivalent, then they belong to the same sign or to equinyms depending on whether one or two interpretations are involved. The representamens in the second row are already less similar to each other. $T\sigma\alpha\rho\lambda\varsigma$ Σ. $\Pi\varepsilon\rho\varsigma$ is the modern Greek spelling of Peirce's name. The number of characters in both representamens is different, thus it might be difficult to establish representamen similarity by a simple mapping of characters. Most likely two interpretations are required in this case. It is even more difficult to detect similarity for the third pair. The 24 numbers for the hours of the right clock correspond to twice the number of hour-lines of the left clock. The minutes of the right clock correspond to the movement of the minute hand of the left clock. There is a representation of movement in both clocks. Thus several aspects of similarity can be established. Finally, the two representamens in the last row only have in common that they are both sets. Otherwise there is no similarity between their representamens. They are still synonyms if their interpretations map them onto the same denotation.

Table 1. Different representamens with similar meanings

1	CHARLES S. PEIRCE	*Charles S. Peirce*	equal signs or equinyms
2	$T\sigma\alpha\rho\lambda\varsigma$ Σ. $\Pi\varepsilon\rho\varsigma$	Charles S. Peirce	(equal signs,) equinyms or strong synonyms
3		23:56	(equal signs,) equinyms or strong synonyms
4	$\{1,2,3\}$	$\{n \in N \mid 1 \leq n \leq 3\}$	strong synonyms

3   How SCA Relates to Other Existing Theories

The majority of other existing semiotic theories appear to be either not formally (mathematically) defined or not triadic. Priss (2015) briefly discusses other formalisations of Peirce's semiotics and approaches to model triadic relations with FCA and argues that these are quite different from SCA. Goguen (1999, p. 1) remarks "Semiotics ... much of the research in this area has been rather vague." His own work is not vague but a formal theory of what he calls Algebraic Semiotics. One difference between SCA and Algebraic Semiotics is that although Goguen discusses some of Peirce's ideas, his main influence was Saussure. Thus his signs are binary and belong to sign systems. His definition of sign systems uses partial orders for sorts and constructors. This is somewhat similar to our use of concept lattices as described below. Goguen then discusses structure preserving morphisms among sign systems which are similar to some of the mappings discussed by Priss (2015) and used in Sects. 5 and 6 of this paper. But another difference between Goguen's work and SCA is that his morphisms are mainly focussed on representamens (in our terminology) and syntactic constructions.

---

[3] A tolerance relation is symmetric and reflexive. An equivalence relation is also transitive.

We would argue that Goguen's method is very useful, but mainly for representamens that are structurally reasonably similar to each other and not for those that are very different. For example, it is fairly straightforward to construct morphisms for the first three rows in Table 1. But for the last row, the only connection between the left and right representamens can be established via their denotations.

Another area that should be mentioned is formal model-theoretic semantics which maps representations into models using interpretations and thus has similar ingredients as SCA. But we would argue that formal semantics is a binary view, and not a triadic one, because their interpretations do not have any structure themselves other than being functions. As far as we are aware, formal semantics is not concerned with questions about the ordering, similarity or quality of interpretations. Formal semantics is mainly concerned with formal languages whereas SCA can be applied to non-formal languages and to questions about how formal languages are used by people as well. In formal semantics it is not possible to discuss the representational and the denotational aspects of a sign separately.

Last but not least, it should be mentioned that semiotics (and SCA) is not the same as usability modelling. While it is possible to ask semiotic questions about how people use signs, one can at the same time ask questions about why certain signs might be used in a certain way based on an analysis of their parts, structures and relations with other signs. Thus usability, semantic and syntactic questions can all be discussed within a single framework in SCA.

4   The Core Definitions of SCA

This section presents the core mathematical definitions of SCA[4].

Definition 1: A *semiotic relation* $S \subseteq I \times R \times D$ is a relation between three sets (a set R of *representamens*, a set D of *denotations* and a set I of *interpretations*) with the condition that any $i \in I$ is a partial function $i : R \twoheadrightarrow D$. A relation instance (i, r, d) with $i(r) = d$ is called a *sign*.

Alternatively, S can be called a set of signs. The sets R, I, D and S need not be disjoint. Thus a denotation, representamen or interpretation can also be a sign itself. It is possible to use total functions instead of partial functions by adding a NULL-element as shown in the next definition. Formalisations involving a NULL-element can be complex because NULL might correspond to negative, missing or contradictory information. A common programming practice is to deliberately check whether a variable is non-NULL before performing an operation that would otherwise crash and to ignore the problem otherwise. Similarly, we will only mention NULL-elements and the fact that the functions are partial in the text below if it is absolutely necessary.

---

[4] A reader who is unfamiliar with FCA could read Sect. 5 first because it contains an example of a concept lattice.

Definition 2: For a semiotic relation S, a *NULL-element* $d_\perp$ is a special kind of denotation with the following conditions: (i) $i(r)$ undefined in $D \Rightarrow i(r) := d_\perp$ in $D \cup \{d_\perp\}$. (ii) $d_\perp \in D \Rightarrow$ all i are total functions.

The following definitions determine the basic structures for each of the three sets. In each case a concept lattice and tolerance relations are defined. Linguists sometimes use the term 'open set' for a set that is large and indeterminate, such as the set of all the words of a natural language. It is feasible to define a tolerance relation on such an open set based on rules. Concept lattices, however, require an explicit set of formal objects. It is therefore advantageous not to incorporate R, I and D directly into concept lattices but to map these sets into concept lattices which model domains for the sets. In applications such mappings might be just partial functions but in that case the sets can be reduced in order to have total functions. Domain lattices can be generated from data or can be preconstructed based on assumptions about the data and then be reused for different applications. Building lattices and defining tolerance relations constitutes by itself some form of interpretation. This can be modelled with SCA as well but that is not further discussed in this paper.

Definition 3: For a set R of representamens, a set $T_R = \{t \mid t \subseteq R \times R\}$ of tolerance relations is defined with a subset $E_R \subseteq T_R$ of equivalence relations. A concept lattice $B(O_R, A_R, J_R)$ called *representamen domain lattice* is defined with sets O_R and A_R, a binary relation $J_R \subseteq O_R \times A_R$ and a function $\beta : R \to B(O_R, A_R, J_R)$ with the condition $\exists_{e \in E_R} \forall_{r_1, r_2 \in R} : (r_1, r_2) \in e \iff \beta(r_1) = \beta(r_2)$.

In applications the condition about mapping equivalent representamens onto the same concept can always be achieved by first constructing β and the lattice and then defining the equivalence relation accordingly. For each tolerance relation on R the function β induces a tolerance relation on the lattice. Ideally a tolerance relation on a lattice should be somehow related to the lattice structure (for example by defining a distance metric on the lattice) but that is a modelling aspect which is not a formal requirement. It should be noted that tolerance relations are expected to be defined on the whole set (i.e., $\forall_{r \in R} : (r, r) \in t$) not just on a subset of R. The next two definitions establish domain lattices for the other sets in a similar manner.

Definition 4: For a set I of interpretations, a set $T_I = \{t \mid t \subseteq I \times I\}$ of tolerance relations is defined with a subset $E_I \subseteq T_I$ of equivalence relations. A concept lattice $B(O_I, A_I, J_I)$ called *interpretation domain lattice* is defined with sets O_I and A_I, a binary relation $J_I \subseteq O_I \times A_I$ and a function $\beta : I \to B(O_I, A_I, J_I)$ with the condition $\exists_{e \in E_I} \forall_{i_1, i_2 \in I} : (i_1, i_2) \in e \iff \beta(i_1) = \beta(i_2)$.

One possibility for defining the lattice is to choose $O_I = I$ and to define $\beta(i)$ as the lowest concept that contains i in its extension. The interpretations could represent who is interpreting (a native speaker, a teacher or a student, a programming language compiler, and so on) and when and where a sign is used. This could involve a containment hierarchy. For example, there could be an interpretation for a whole book, with separate interpretations for chapters and paragraphs. It is possible to combine all of these containment orders into one concept lattice because using the method of Dedekind closure any partial order can be embedded into a lattice. Once such a lattice has been formed, one can then set the set A_I to correspond to the meet-irreducible lattice elements. But this is just one possibility for constructing the lattice. It could also be constructed in a totally different manner for other applications.

Definition 5: For a set D of denotations, a set $T_D = \{t \mid t \subseteq D \times D\}$ of tolerance relations is defined with a subset $E_D \subseteq T_D$ of equivalence relations with the condition $(d, d_\perp) \in e \in E_D \Rightarrow d = d_\perp$. A concept lattice $B(O_D, A_D, J_D)$ called *denotation domain lattice* is defined with sets O_D and A_D, a binary relation $J_D \subseteq O_D \times A_D$ and a function $\beta : D \to B(O_D, A_D, J_D)$ with the conditions that if $\beta(d_\perp)$ exists it is the bottom element of the lattice and $\exists_{e \in E_D} \forall_{d_1, d_2 \in D} :$ $(d_1, d_2) \in e \iff \beta(d_1) = \beta(d_2)$.

A denotation domain lattice represents the denotational knowledge of a domain. It could be derived from data or from an ontology (or textbook knowledge in an educational application) using any of the usual FCA techniques for encoding knowledge. The denotational knowledge could also be provided using other knowledge representation techniques (such as conceptual graphs, description logic or formal ontologies). But for the purposes of SCA, one would then need to extract lattices from such knowledge. The next definition shows how some common linguistic terms are formalised in SCA. These definitions only use the relations from Definitions 3–5, not the concept lattices. Thus they would still be applicable if some formalisation other than lattices was used.

Definition 6: For a semiotic relation S with $t \in T_D$, $I_1 \subseteq I$, $e \in E_R$, $e_D \in E_D$,

(a) I_1 is *e-compatible* $\Leftrightarrow \forall_{(r_1,r_2) \in e, i_1, i_2 \in I_1, i_1(r_1) \neq d_\perp, i_2(r_2) \neq d_\perp} : (i_1(r_1), i_2(r_2)) \in t$
(b) I_1 is *e-mergeable* $\Leftrightarrow \forall_{(r_1,r_2) \in e, i_1, i_2 \in I_1, i_1(r_1) \neq d_\perp, i_2(r_2) \neq d_\perp} : i_1(r_1) = i_2(r_2)$
(c) $(i_1, r_1, d_1) = (i_2, r_2, d_2) \Leftrightarrow i_1 = i_2, (r_1, r_2) \in e, d_1 = d_2$.
(d) (i_1, r_1, d_1) and (i_2, r_2, d_2) are *strong synonyms* $\Leftrightarrow (r_1, r_2) \notin e$ and $(d_1, d_2) \in e_D$
(e) (i_1, r_1, d_1) and (i_2, r_2, d_2) are *equinyms* $\Leftrightarrow (r_1, r_2) \in e$ and $(d_1, d_2) \in e_D$
(f) (i_1, r_1, d_1) and (i_2, r_2, d_2) are *synonyms* $\Leftrightarrow (r_1, r_2) \notin e$ and $(d_1, d_2) \in t$
(g) (i_1, r_1, d_1) and (i_2, r_2, d_2) are *polysemous* $\Leftrightarrow (r_1, r_2) \in e$ and $(d_1, d_2) \in t$
(h) (i_1, r_1, d_1) and (i_2, r_2, d_2) are *homographs* $\Leftrightarrow (r_1, r_2) \in e$ and $(d_1, d_2) \notin t$
(i) (i, r, d) is *anonymous* $\Leftrightarrow r = d$

If e is clear from the context, the prefix 'e-' can be omitted. The notions in (d) to (h) depend on i_1 and i_2, thus in cases of possible ambiguity, one could write 'i_1, i_2-synonyms' and so on. Mergeability means that the interpretations can be merged because the result of the merger is still a function. Representamens as physical manifestations usually display minute variations. For example, two spoken words or two handwritten words are probably never totally equal. Even two computerised images that look the same may not be totally equal if one of them has been compressed or encoded differently. Therefore we allow for two equal signs to have equivalent instead of equal representamens. If $i \in I$ is not e-mergeable with itself, either e could be changed by reducing the size of the equivalence classes or I could be changed by splitting interpretations which are not e-mergeable with themselves.

As mentioned before, it is quite restrictive to require two equal signs to have the same interpretation. Therefore the other notions from Definition 6 describe forms of similarity among interpretations and among signs which are weaker than equality. Synonymy, polysemy and homographs are formalisations of the usual linguistic notions. Equinymy was coined by Priss (2004) and refers to the same representamen being used with the same meaning under different interpretations. Equinymy probably expresses what one might intuitively think it means for signs to be 'the same'. An example of anonymous signs are literals in programming languages. Priss (2004) argues that mathematical variables are anonymous signs as elaborated below.

Proposition 1: For a semiotic relation S with $t \in T_D$, $e \in E_R$, $e_D \in E_D$, $i_1, i_2 \in I$

(a) e-compatibility and e-mergeability are tolerance relations on $I \times I$. If $d_{\perp}$ does not exist, then e-mergeability is an equivalence relation on $I \times I$.
(b) For given, fixed i_1 and i_2, equinymy is an equivalence relation, synonymy, strong synonymy, polysemy and homographs are tolerance relations on $S \times S$.
(c) Strong synonymy implies synonymy. Equinymy implies polysemy.
(d) Compatible interpretations are free of homographs.
(e) For mergeable interpretations, polysemy and equinymy are the same.
(f) Two anonymous signs are equinyms if their denotations are equal. If $e = e_D$ or e_D is the identity relation, two anonymous signs cannot be strong synonyms. If $t = e$ or t is the identity relation, they cannot be synonyms.
(g) If $t = e = e_D$ or t and e_D are identity relations, then two anonymous signs in mergeable interpretations can only be equal (if $i_1 = i_2$), equinyms (if $i_1 \neq i_2$) or not equal. They cannot be synonyms or homographs.

The statements in Proposition 1 follow directly from Definition 6. It should be noted that compatibility and mergeability need not be elements of T_I. The set T_I contains those tolerance relations which are explicitly defined for I, not necessarily those which are emerging based on other defined structures. Statement (g) describes how variables are normally used in mathematics. Mathematical texts usually only involve one interpretation which means signs are either equal or not equal. In mathematics, variable names are not distinguishable from their content and equivalence among variable names corresponds to denotational equality. In programming languages, however, signs are not anonymous and interpretations are not always mergeable. We conclude this section with the definition of SCA:

Definition 7: A semiotic relation with concept lattices as presented in Definitions 3–5 is called a *semiotic system*. The study of semiotic systems is called a *semiotic-conceptual analysis*.

A next step is to investigate functions between the concept lattices. Because the interpretations are partial functions from representamens into denotations this leads to the question as to whether they induce partial functions from the representamen domain lattice to the denotation domain lattice. Priss (2015) discusses some conditions for such functions which we will omit in this paper because we believe that further applications are needed in order to determine what is most promising. The following sections show some examples of using such functions in educational applications.

5 An Example: Reading Hasse Diagrams of Concept Lattices

As a first example we are investigating challenges involved in teaching students about concept lattices and their representations as Hasse diagrams. Figure 1 shows a formal context and a Hasse diagram of a concept lattice. A Hasse diagram is a representation of a partially ordered set which has the elements as nodes (here nodes 0, 1, ... 4) and their immediate relationships as edges. The edges in the diagram are directed. This means that going up in the diagram corresponds to going up in the partially ordered set. Because the ordering in a partially ordered set is transitive, node 4 is not only below node 3 (its immediate neighbour) but also below all the other nodes. Node 0 is above all other nodes. Transitivity is implied but not explicitly represented in a Hasse diagram because there are no lines between node 4 and node 2 and so on. The implied transitivity needs to be explained to someone who does not know what a Hasse diagram is.

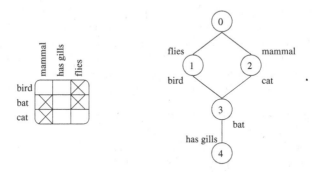

Fig. 1. A formal context and a Hasse diagram of a concept lattice

The Hasse diagram in Fig. 1 shows a concept lattice which means that every set of nodes has a unique supremum and a unique infimum and it corresponds to a formal context as displayed in the left-hand side of the figure. The objects are written slightly below the nodes (or concepts) they belong to and the attributes slightly above. In this example the objects are bird, cat and bat and the attributes are flies, mammal and has gills. A concept in FCA has an extension and an intension. An extension consists of all objects from all concepts below a concept. An intension is defined analogously. In this example the concept of node 3 has bat in its extension and flies and mammal in its intension. The theories of lattices in general and of FCA in specific are well developed. Therefore the body of knowledge that is connected with this small lattice is larger than one might think when one first sees this little figure which consists of just a few nodes, edges and labels.

In our experience with showing FCA to students (and users in general) there are a number of typical questions that arise when they first see Hasse diagrams:

- What is the purpose of the top and bottom node?
- Why are there unlabelled nodes?
- How can the extensions and intensions be read from the diagram?
- What is the relationship between nodes that do not have an edge between them but can be reached via a path?
- What is a supremum or an infimum?
- How can one tell whether it is a lattice?

Figure 2 shows a modelling of Hasse diagrams with respect to a representamen domain lattice (on the left) and of notions from lattice theory as a denotation domain lattice on the right. In both lattices only the representamens and denotations are shown but not the formal objects and attributes. For the representamen lattice, it is assumed that for a representamen r there is a set $r^{J_R} \subseteq A_R$ of attributes which is assigned to r. The set r^{J_R} need not be an intension of $B(O_R, A_R, J_R)$. Then $\beta(r)$ is defined as the largest concept that contains r^{J_R} in its intension. The definition of $\beta(d)$ is analogously. Building representamen and denotation domain lattices involves modelling. Thus the lattices in Fig. 2 are not to be understood as 'ultimate truth' but instead as a teacher's model. The goal of this is example is not to discuss whether these lattices are correct or not but whether building and analysing such lattices can provide insights about a semiotic relation.

The superconcept ordering in both lattices corresponds to prerequisite knowledge for a student. For example one needs to know what operators and sets are before one can learn what a relation is. The dashed lines in Fig. 2 show the result of an interpretation for some of the representamens. Every concept of the denotation domain lattice represents a different meaning. Because this example comes from a mathematical domain, it may be sufficient to assume a single interpretation. Figure 2 shows that the structures between the representamen and denotation domain lattices are quite different. In this example this means that Hasse diagrams are structurally quite different from lattices. There are connections between some of the core notions, such as 'node' representing 'concept' and 'edge' representing immediate neighbours in the ordering relation. Some of the representamens represent activities on the Hasse diagram: traversing edges up or down and collecting labels on the way represents extensions and intensions. The node/edge relationship is used for identifying irreducible concepts by counting the number of edges that lead to a node from above or from below.

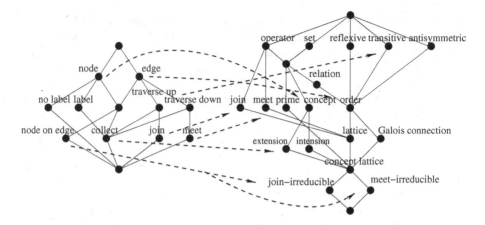

Fig. 2. A representamen and a denotation domain lattice

Representamens such as labels for objects and attributes do not seem important for the denotation domain lattice, presumably because mathematically they are just elements of sets. The relationships between 'edge' and 'traversal' and 'transitive' and 'order' are inverse. For diagrams one must first know what an edge is before one can talk about traversing edges. But in order to define an order relation one must first know what transitivity is.

In order to address the questions students have when they first see Hasse diagrams of concept lattices, a diagram such as Fig. 2 helps to investigate where exactly their misconceptions are coming from. For example, a question about the purpose of the top and bottom node indicates that the student does not know what a complete lattice is. Using the denotation domain lattice in Fig. 2 a teacher could ask a student about order, join and meet because these are prerequisite to understanding lattices. If the student also has problems with these concepts, then the teacher could move up further in the lattice. It should be mentioned that this approach of representing an ordered set of prerequisite knowledge is also used in Knowledge Space Theory (Albert and Lukas 1999; Falmagne et al. 2013). Connections between Knowledge Space Theory and FCA are well known but are beyond the scope of this paper.

6   An Example: The Meaning of the Equality Sign

Prediger (2010) discusses how students successively enhance their understanding of the equality sign. At first in primary school, students interpret the equals sign as a request to calculate something (such as '$2 + 3 = ?$'). Prediger calls this the operational use because it is a request for performing an operation. Later, students learn a 'relational' meaning of the equals sign which could involve symmetric identities ($4+5 = 5+4$), general equivalences ($(a-b)(a+b) = a^2-b^2$), searching for unknowns ($x^2 = 6 - x$) and contextual uses ($a^2 + b^2 = c^2$) where the variables are meaningful in a context, such as characterising a right-angled triangle. A further, different case are specification uses (such as defining $x := 4$).

Priss et al. (2012) model the denotational content of the equals sign and other equation, assignment and comparison operators using FCA. Their resulting concept lattice is presented in the right-hand side of Fig. 3. We are now revisiting this example by building a representamen domain lattice (left-hand side of Fig. 3) and investigating interpretation-induced partial functions. The left lattice is constructed with $O_R = R$ and $\beta(r)$ is the object concept of r. For the right lattice, $\beta(d)$ is constructed as in the previous example. The denotation domain lattice is built using formal objects which are examples of uses of the equals sign, inequality ($>$), equivalence ($\Leftrightarrow$), basic operations from programming languages: not-equal ($!=$), test for equality ($==$) and Boolean operators ($\&\&$). It should be noted that for the examples with more than one operator, the main operator ($\Leftrightarrow$, $==$ and $\&\&$) is the one that is investigated. The formal attributes are 'operation', 'contextual', 'definition' (i.e., specification) and 'law' (i.e., equivalence), 'test' and whether the statements are true for all values of the variables or just for some. A 'definition' for other symbols than '=' defines a set of possible values for a variable (e.g., $i > 1$). A 'test' is a request to evaluate an expression with respect to variables with given values. The representamen domain lattice classifies the different operation symbols with respect to their parts and their complexity. Even though '$\Leftrightarrow$' could be one character, we argue that one could see it as two arrows and thus as more complex. As for the previous example, the lattices are just an example of modelling by one teacher for the purpose of evaluating student progress.

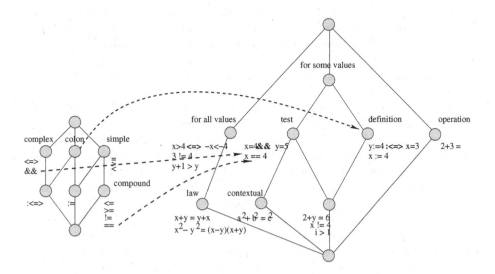

Fig. 3. Equation, assignment and comparison operators

The dashed lines in Fig. 3 show the result of an interpretation which maps representamens into denotations. A single representamen can be mapped onto the concept in the denotation domain lattice that is the join concept of all its different denotations or a concept of the representamen lattice can be mapped depending on its extension. If a representamen is used for different denotations this means that different interpretations have to be involved. In this example, only == and && are used for exactly one denotation ('test'). Furthermore, the representamens in the extension of the concept labelled 'colon' are only used with one denotation ('definition'). These interpretation instances are shown by the dashed lines. All other representamens and all other extensions of representamens are used for different denotations whose join is the top concept in the denotation domain lattice. These interpretation instances are not drawn in Fig. 3. Overall there is not much structure from the representamen domain lattice that is preserved in the denotation domain lattice. Many representamens are homographs because the join of their images under different interpretations is the top concept of the denotation domain lattice.

In Fig. 4 the denotation domain lattice has been restructured in order to support more interpretation instances which do not point towards the top concept. It should be noted that the lattice on the right-hand side is shown as an incomplete 'nested line diagram' which has nodes missing. The restructuring of the lattice has also used ideas from APOS Theory (Dubinsky and Mcdonald 2002) according to which learning of mathematical concepts often progresses from action- to progress- to object-level understanding. At an action level understanding students perform operations without really knowing much about them. Prediger's (2010) operational use of the equals sign appears to be action level. With respect to the representamen domain lattice it appears reasonable to separate the interpretation images of complex and simple representamens. From an APOS viewpoint this could correspond to a difference between process and object level understanding. At an object level, a concept becomes itself reified and part of another concept. In the examples with more than one operator, the simple operators become objectified. For example in $x = 4$ && $y = 5$, the Boolean operator && is primary whereas its left and right operands are only evaluated with respect to their truth values.

Altogether the denotation domain lattice in Fig. 4 might be an example of what is called a 'genetic decomposition' in APOS Theory. A student's conceptual learning should move from the top of the lattice to the bottom. A complete understanding of the comparison operators is achieved when a student knows that the operators are used with different meanings and knows exactly what each operator means in the context it is represented in. Depending on what a student says about one of the operators and in particular depending on what kinds of errors a student makes when using an operator, the student's conceptual stage at that point could be pinpointed in the denotation domain lattice.

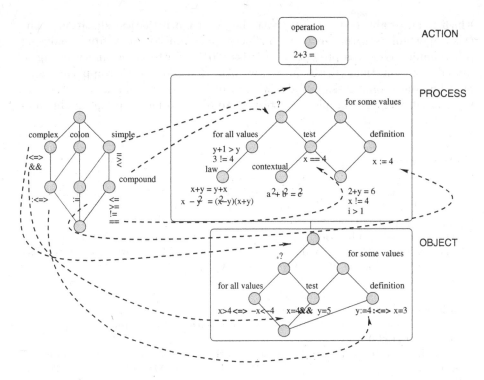

Fig. 4. Remodelling of the denotation domain lattice from Fig. 3

7   Conclusion

This paper shows how the modelling of some of the concepts of a domain in combination and in contrast with a modelling of the representamens that are used for the domain can serve as a tool for analysing the knowledge state a student is in. If a student uses terms incorrectly then the student's interpretation must be different from the teacher's interpretation. From a theoretical viewpoint, mapping the signs uttered by a student into a denotation domain lattice shows possible misconceptions on the student's part. From a practical viewpoint, this implies that it is important to observe how a student talks about denotations while he or she performs tasks within the domain. For example, a student could be asked to perform a mathematical calculation and discuss it at the same time. In that manner any misalignment between representamens and denotations might become apparent for the teacher. If a student just writes a textual exam or just performs a calculation, it could be possible that the student just reproduces material he or she has memorised without actually understanding it.

The examples in the previous sections show methods that use denotation domain lattices for representing prerequisite conditions within domain knowledge similar to Knowledge Space Theory. Furthermore, a denotation domain lattice is shown that could serve as a genetic decomposition in the sense of APOS Theory which is an established constructivist theory of mathematics education. Thus this paper enlarges the area of possible applications of SCA from analysing programming code (as presented by Priss (2015)) to teaching and learning of formal representations. In the future, we plan to explore further applications of SCA for educational tasks but also with respect to structural analysis of formal representations, for example, with respect to object-oriented programming and XML.

References

Albert, D., Lukas, J. (eds.): Knowledge Spaces: Theories, Empirical Research, Applications. Lawrence Erlbaum Associates, Mahwah (1999)

Dubinsky, E., Mcdonald, M.A.: APOS: a constructivist theory of learning in undergraduate mathematics education research. In: Holton, D., Artigue, M., Kirchgräber, U., Hillel, J., Niss, M., Schoenfeld, A. (eds.) The Teaching and Learning of Mathematics at University Level. ICMI Study Series, vol. 7, pp. 275–282. Springer, Dordrecht (2002)

Falmagne, J.-C., Albert, D., Doble, D., Eppstein, D., Hu, X.: Knowledge Spaces: Applications in Education. Springer, Heidelberg (2013)

Ganter, B., Wille, R.: Formal Concept Analysis: Mathematical Foundations. Springer, Heidelberg (1999)

Goguen, J.: An introduction to algebraic semiotics, with application to user interface design. In: Nehaniv, C.L. (ed.) CMAA 1998. LNCS (LNAI), vol. 1562, pp. 242–291. Springer, Heidelberg (1999)

Prediger, S.: How to develop mathematics-for-teaching and for understanding: the case of meanings of the equal sign. J. Math. Teach. Educ. **13**, 73–93 (2010)

Priss, U.: Relational concept analysis: semantic structures in dictionaries and lexical databases. Ph.D. thesis, Verlag Shaker, Aachen (1998)

Priss, U.: Signs and formal concepts. In: Eklund, P. (ed.) ICFCA 2004. LNCS (LNAI), vol. 2961, pp. 28–38. Springer, Heidelberg (2004)

Priss, U., Riegler, P., Jensen, N.: Using FCA for modelling conceptual difficulties in learning processes. In: Domenach, F., Ignatov, D., Poelmans, J. (eds.) Contributions to the 10th International Conference on Formal Concept Analysis (ICFCA 2012), pp. 161–173 (2012)

Priss, U.: An introduction to semiotic-conceptual analysis with formal concept analysis. In: Yahia, S., Konecny, J. (eds.) Proceedings of the Twelfth International Conference on Concept Lattices and Their Applications, Clermont-Ferrand, pp. 135–146 (2015)

Organised Crime and Social Media: Detecting and Corroborating Weak Signals of Human Trafficking Online

Simon Andrews[1,2](✉), Ben Brewster[2], and Tony Day[2]

[1] Conceptual Structures Research Group, Sheffield Hallam University, Sheffield, UK
[2] Centre of Excellence in Terrorism, Resilience,
Intelligence and Organised Crime Research,
Communication and Computing Research Centre,
Sheffield Hallam University, Sheffield, UK
{s.andrews,b.brewster,t.day}@shu.ac.uk

Abstract. This paper describes an approach for detecting the presence or emergence of Organised Crime (OC) signals on Social Media. It shows how words and phrases, used by members of the public in Social Media, can be treated as weak signals of OC, enabling information to be classified according to a taxonomy of OC. Formal Concept Analysis is used to group information sources, according to Crime and Location, thus providing a means of corroboration and creating OC Concepts that can be used to alert police analysts to the possible presence of OC. The analyst is able to 'drill down' into an OC Concept of interest, discovering additional information that may be pertinent to the crime. The paper describes the implementation of this approach into a fully-functional prototype software system, incorporating a Social Media Scanning System and a map-based user interface. The approach and system are illustrated using the Trafficking of Human Beings as an example. Real data is used to obtain results that show that weak signals of OC have been detected and corroborated, thus alerting to the possible presence of OC.

1 Introduction

The vociferous proliferation of the Internet, and more recently Social Media, into society and the everyday lives of its citizens has, over the last fifteen or so years, resulted in a sea-change in the behaviours and perceptions we have in relation to the information that is shared freely online [12]. Such behaviour has resulted in the creation of a vast repository of information that holds potential value for police investigations, and the emergence of the open-source researcher as a valuable skill-set within the analytical repertoire of the police and security agencies. Resources such as social media, RSS news feeds, interactive street-maps and online directory services all provide valuable stores of information that can be used to support existing investigative and analytical practices in response to serious and organised crime. This paper focuses on the identification, extraction and corroboration of data from social media using automated data acquisition,

O. Haemmerlé et al. (Eds.): ICCS 2016, LNAI 9717, pp. 137–150, 2016.
DOI: 10.1007/978-3-319-40985-6_11

natural language processing and formal concept analysis (FCA), specifically in order to identify what we will refer to as 'weak signals' of human trafficking, and to transform these signals into corroborated alerts linked to the presence or emergence of human trafficking activity.

The concept of weak signals is abstracted from the Canadian Criminal Intelligence Service's (CISC) definitions of primary and secondary indicators [14], and the perception that in reality there is little tangible value to be extracted from isolated indicators as there is potential for them to be indicative of a variety of phenomena. However, when these indicators are grouped under certain conditions, such as proximity to a certain location and type of activity, they can begin to provide insights into to the presence or emergence of crime. It is with this definition, and the notion of 'weak signals' that we use as the basis of this paper and the approached presented within it.

Perhaps the greatest shift in the use of the internet over the last 10–15 years or so is the relative phenomenon which is the usage of Social Media among normal citizens. In the aftermath of the events which followed the killing of Mark Duggan in 2011, a HMIC commissioned review highlighted significant inefficiencies in the way that authorities were equipped to deal with social media as an intelligence source [9]. Social media intelligence, or 'SOCMINT', provides opportunities for providing insights into events and groups, enhancing situational awareness, and enabling the identification of criminal intent [10].

In order to demonstrate the potentially utility of SOCMINT, in respect of identifying the presence and/or emergence of organised crime, the problem domain of Human Trafficking will be used as the exemplar throughout this paper. Human trafficking operates on a vast scale, impacting upon almost every country in the world as an origin, transit or destination location for the movement and exploitation of human beings. Trafficking is so defined by article 3 of the Palermo protocol as the 'recruitment, transportation, transfer, harbouring or receipt of persons, by means of the threat or use of force or other forms of coercion, of abduction, of fraud, of deception, of the abuse of power or of a position of vulnerability or of the giving or receiving of payments or benefits to achieve the consent of a person having control over another person, for the purpose of exploitation' [15].

Europol [5] has in the past acknowledged the growing criminal dependence on the internet and the increasingly trans-European perspective of serious and organised crime. These changes in the way that information is created and shared, combined with the diversity in the way that existing forms of criminality are being conducted provides the opportunity, and desire, for the development of new means to assist law enforcement in combating such crime. To provide one such approach to enable this, the tools described here facilitate the identification, extraction, processing, analysis and presentation of data from open sources, such as social media, that can provide insight into the emergence and presence of crime. While at one end of the scale international intelligence agencies such as the CIA's PRISM programme are facillitating the aquisitoin, fusion and analysis of vast amounts of data from disperate sources, the (known) resources and capability of law enforcement agencies (both locally, regionally and even internationally) are much more modest with the use of data from open sources and social media often a manual task, and the remit of just a few specialist analysts and officers within each force.

2   Taxonomy of Organised Crime

In beginning to model its constituent elements it is necessary to ascertain a thorough understanding of the actual problem domain - human trafficking, by drawing upon established definitions used to describe and diagnose the problem by the practitioner base. The UNODC [15] have defined, using the UN Palermo protocol as the basis, what they refer to as, the three constituent 'elements' of trafficking, these being the 'act', 'means' and 'purpose', see Fig. 1. Firstly, the 'Act' refers to what is being done, this can include context such as whether and how the victim has been recruited, transported, transferred or harboured. The question of how this is being achieved is answered by the 'Means' which seeks to establish whether force is being used as the basis of manipulation, such as through kidnapping, abduction or the exploitation of vulnerabilities, or more subtle methods such as through fraud, imposing financial dependencies or coercion. The final element, the 'Purpose' establishes the reason why the act and means are taking place, or to put it simply - the form of exploitation behind the act and means, be it for forced labour, sexual exploitation and prostitution, organ harvesting or domestic servitude.

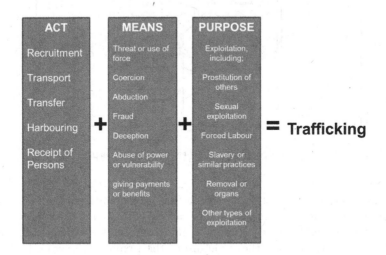

Fig. 1. The elements of Human Trafficking

This definition and categorisation provides an ideal underpinning for the formation of a taxonomy of Human Trafficking that can be used to form the basis of an approach to automatically identify and extract valuable data from open sources (Fig. 2). This taxonomy in actuality forms part of a larger model, consisting of elements of a broader range of organised crime threats, including the cultivation and distribution of illegal narcotics. The elements of the taxonomy are defined across for types, visualised here using vertical lanes. The first of these four lanes starting on the left is used as a high level categorisation used to separate between different crime types. The second lane deals with different elements within a specific type of crime - in this case one of the three component parts of trafficking, while subsequent lanes, type and element, are used to show further more specialised aspects of these elements.

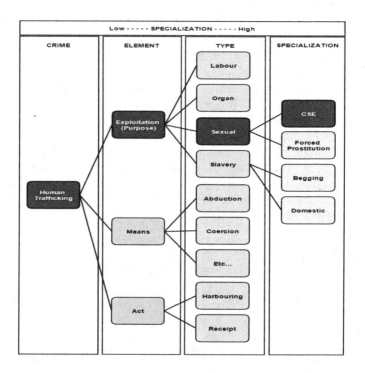

Fig. 2. Organised crime taxonomy (excerpt)

Each of the nodes contained within the taxonomy represents a ruleset designed to determine content's, in this case a particular twitter posting, relevance to the subject matter - Human Trafficking. The level of specialisation of the rules themselves follows the structure of the taxonomy, moving from more generic words or phrases that may indicate Human Trafficking used at the higher levels of the taxonomy, where as as the more specialised end of the taxonomy, more nuanced rules that may allude to the presence of criminality are used.

3   Weak Signals of Organised Crime

To enable the development of a taxonomy to begin to model and structure the information deemed useful to extract we can refer to a wealth of literature from both academic and practitioner perspectives that provide insights into the factors that contribute to and indicate organised crime. These indicators vary from high level, secondary information such as Political, Economic, Socio-cultural, Technological, Legal and Environmental (PESTLE) factors, right down to operationally oriented information that helps us to identify potential victims of trafficking. Existing models to anticipate changes and developments in organised criminality accross geographic areas have focused on this kind of data alongside existing crime statistics [18].

In the past, and to some extent a problem that still exists, a lack of information and common understanding about what human trafficking is has hindered the impact and effectiveness of efforts to combat it [17]. Despite varied and wide-ranging counter-trafficking initiatives from NGO's, Law Enforcement and Governments, reliable information regarding the magnitude and nature of trafficking across regional and national borders is still hard to come by due to a number of issues around the sharing, fusion and understanding of data that is already being collected [8]. The purpose of the approach developed and described in this paper is. not to provide a statistically accurate representation of the presence and emergence of trafficking but rather to increase the access and usability of data from previously untapped open-sources. In previous work, we have discussed indicators across a three-level model [2] moving from credible and accepted indicators of trafficking at level 1 of the model, through to the observations and content created online, including on social media, by citizens regarding these 'weak signals'.

In this paper we discuss the latter and more specifically the modelling and use of this information as 'weak signals' that allude to the presence and/or emergence of criminality in citizen generated content, whilst using the formal definitions and doctrine that exists to underpin the framework and organisation of the model itself. Perhaps the most comprehensive list of indicators comes from the UNODC [16] provide an extensive list of indicators categorised by different types of exploitation such as; domestic servitude, child, sexual exploitation, labour exploitation and begging/petty crime - the labels used in the taxonomy structure outlined in Fig. 2. Using sexual exploitation as an example indicators include things such as the appearance that persons are under the specific control of another, that the person(s) appear to own little clothing, appear to rely on their employer for basic amenities, transport and accommodation and more. Although in this form, these indicators are quite abstract and it can may be potentially difficult to see how they may manifest in real, open-source, data - it is possible to develop rules looking for keywords and phrases that can provide 'weak-signals' of their existence online.

In order to facilitate the identification and extraction of these weak signals in social media and other open-sources, what we will refer to as 'contextual extraction' methods [3] are used in order to identify, and subsequently extract, key entities and facts (i.e. previously unknown relationships between different entities) from the data. This approach to information extraction using natural-language processing builds upon the existing principles of template based information extraction [13], also sometimes referred to as 'Atomic Fact Extraction' [3] These 'facts' enable the extraction of entities within a specific context, i.e. locations in relation to an arrest or type of exploitation on a per sentence basis.

While in isolation, the extraction of these entities on their own does not necessarily provide much actionable information, it is possible using rules that attempt to infer relations between them to begin to make some assumptions about the data and its content. For example a single tweet may contain multiple locations and other entities, but without some means to establish a relationship between the two there is no way to automatically infer, with any confidence at least, that they are linked. Fortunately, through the use of contextual extraction, and the aforementioned 'facts' we can infer these relationships in a number of ways, using prepositions, parts-of-speech tagging and Boolean operators that specify distance between words and other parameters. As the examples in discussed in this paper refer to data from Twitter only, these relationships are done on a 'per sentence' basis. The following example shows how this works in practice: From these rules we can begin to make some assumptions about the entities being extracted. For example, it is now possible to ascertain with a degree of confidence that specific locations are in reference to a specific event. At this point, it is important to acknowledge the challenges posed by the use of SMS-language (textese) as communication via services such as twitter do not necessarily adhere to strict grammer or syntax conventions. Although a number of novel approaches to handle this type of language are in development (see, for example [11]) due to the use of examples that use accepted, formal terminology, we do not address this issue here (Fig. 3).

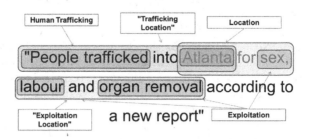

Fig. 3. Entity and fact extraction in social media

4   Categorisation and Filtering

In addition to extracting facts and entities from the input data, similarly techniques are also utilised as the basis of a rule based approach to classify content against a number of pre-defined categories. Membership to one or more of these categories is then used as the basis for content filtering. If the content analysed by the crawler does not meet the criteria of at least one of the rules within the parameters defined in the content categorisation process, it is then disregarded. The categorisation model is defined using a similar approach to the entity and fact extraction model, utilising a number of 'hand-crafted' rules organised in a hierarchical structure, with the only key difference being that rather than being designed to identify and extract specific pieces of data and/or information they aim to discern the relevance of the content against the defined topics using the same taxonomy structure defined in Fig. 2. The rules themselves use a range of techniques, again focusing on the identification of keywords and phrases. A number of examples of the phrases and keywords used as part of the categorisation taxonomy are shown in Table 1.

Table 1. Weak signals - keywords and phrases

Weak signal	Keywords and phrases
Physical injury	Subjected to violence
	Timid
	Forced to have sex
	Women beaten
Physical appearance	Provocative dress
	Live with a group of women
	Unhappy
Unable to leave place of residence	Afraid to leave
	Under control of others
	Financially dependant
Irregular movement of individuals	Men come and go at all hours
	Women do not appear to leave
	Lots of activity at night

5   A Social Media Scanning System

To implement the content extraction and categorisation models, an integrated pipeline that facilitates the crawling of social media is put in place. This process manages and enables the seamless collection, restructuring, processing, filtering and output of the data in preparation for further analysis. The stages of the data preparation and processing pipeline is shown in Fig. 4.

Utilising the 'Search API' offered by Twitter [7], queries can be made against the service's index of recent and/or popular posts from the previous seven days, with only the most relevant tweets returned from during the time period. At the time of writing, the amount of data returned is limited by the API's rate limit, current set at 180 queries per 15 min. This amount is subject to change. In terms of the queries themselves, a number of pre-defined operators exist that allow for the matching of keywords, exact phrases and other operations.

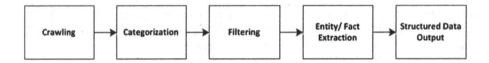

Fig. 4. Social media scanning process

6   Corroborating Information Using Formal Concept Analysis

A single Tweet containing a weak signal of OC is not a sensible basis for Law Enforcement Agencies (LEAs) to take action. However, if a number of sources contain weak signals of the same element of OC from the same location, then this may form a credible basis to warrant further investigation. Such corroboration can be automated by the application of Formal Concept Analysis (FCA) [6] to the structured data extracted from the information sources. FCA can be used to cluster the sources into so-called *OC Threat Concepts* (or just *OC Concepts*) where one shared attribute is a location and another shared attribute is an element of OC. When mining the data for formal concepts, if a minimum support is set, say 10 information sources, only OC Threat Concepts with a least 10 sources will be obtained.

To carry out FCA, the structured data extracted from the information sources must be scaled into a formal context. For example, each location in the data becomes a formal attribute in the context. The weak signals of elements of OC are scaled using the Taxonomy. So, for example, a weak signal may indicate Human Trafficking, and thus the source in which the signal was contained will be labeled with the formal attribute *Crime-HumanTrafficking*. Several different weak signals may all point to Human Trafficking, and thus sources containing any of them would all be labeled with the attribute *Crime-HumanTrafficking*. Other weak signals may point to more specific elements of Human Trafficking, such as Exploitation which is a component of Human Trafficking (from the taxonomy). A source with such a weak signal will be labeled with both *Crime-HumanTrafficking* and *Element-Exploitation*. Thus the general 'is a part of' rule in a taxonomy becomes naturally scaled in FCA (see Fig. 5).

OC Taxonomy	Human Trafficking	Exploitation	Sexual	CSE
weak signal of THB	×			
weak signal of Exploitation	×	×		
weak signal of Sexual Exploitation	×	×	×	
weak signal of CSE	×	×	×	×

Fig. 5. A formal context scaling part of the OC taxonomy (CSE is Child Sexual Exploitation)

Using a data set created from 29096 Tweets as information sources, obtained by scanning for Tweets containing weak signals of OC, a formal context was created by scaling the extracted structured data as above. Using a minimum support of 80, the context was mined for OC Threat Concepts using a modification of the In-Close concept miner [1]. The result is visualised as a formal concept tree in Fig. 6.

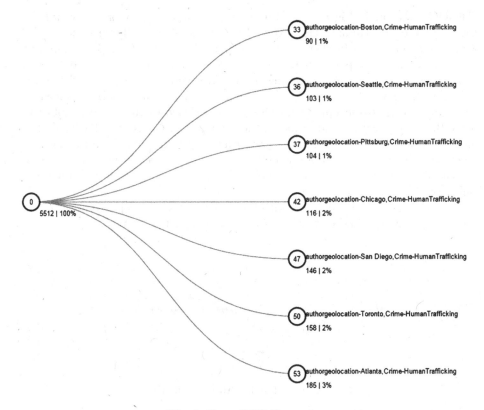

Fig. 6. Tree of OC Concepts

In the tree, the head node is the concept containing all the Tweets from concepts that satisfy the minimum support (5512 Tweets) and each of the branches is to an OC Concept - a concept where one attribute is a location and another is an OC. In this example, every OC is Human Trafficking as this was the type of OC being searched for by the Scanning System. The number inside each node is simply a concept ID number assigned by the concept miner. The number outside the node, below the list of attributes, is the object count (the number of Tweets contained in the concept) and in each case this is above the minimum support threshold of 80. Thus concept 53, for example, has the attributes *authorlocation-Atlanta* and *Crime-HumanTrafficking*, and has 185 objects. In other words, within the data set there are 185 Tweets that have the author location Atlanta and contain a weak signal of Human Trafficking. With this high level of corroboration, a police analyst will be alerted to investigate this further, and a possible next step in the investigation is automated by FCA in the form of a 'drill-down' to the OC Concept's sub-concepts.

6.1   OC Concept Drill-Down

The OC Concepts in Fig. 6 contain limited information - they only have a location and the OC Human Trafficking. However, individual Tweets in the OC Concept may contain further information pertinent to the OC. But examining 185 Tweets, for example, although far less work than examining 29096 Tweets, is nonetheless quite time consuming. However, several Tweets in the OC Concept may all share the same additional information and this can be divulged by examining the sub-concepts of an OC Concept. Each of the sub-concepts will have the same location and crime as the original OC Concept but with one or more additional attributes from the structured data extracted from the Tweets. Such a result can easily be obtained by mining the data for concepts that contain the attributes of the OC Concept and at the same time reducing the minimum support required.

Figure 7 shows a concept tree with the 'Atlanta' OC Concept from Fig. 6 and its sub-concepts produced when the minimum support is set to 5.

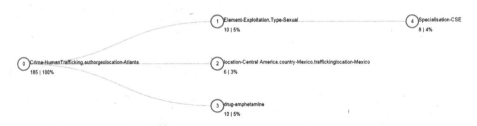

Fig. 7. Drill-Down for the 'Atlanta' OC Concept

In the tree, concept 3 shows that 10 of the 185 'Atlanta' Tweets also contain a reference to the drug amphetamine. They may not all contain the actual word *amphetamine*, but they will all contain a word or phrase that is commonly used to mean or refer to amphetamine. But using lists of such words and phrases, the entity extraction process carried to produce the structured data will thus label each of these Tweets with the attribute *drug-amphetamine*, which in turn enables the FCA to group them together.

Concept 2 shows that 6 of the 185 Atlanta Tweets also contain the location Central America and the county Mexico. Furthermore, a semantic rule in the entity extraction process has determined that Mexico is being referred to in the Tweet as a trafficking location and thus these Tweets are labeled with the attribute *traffickinglocation-Mexico*.

Concept 1 shows that 10 of the 185 Atlanta Tweets contain weak signals of the OC Human Trafficking *element* Exploitation and the *exploitation type* Sexual. Furthermore, in 8 of those 10 Tweets there are weak signals of CSE, further specialising the OC.

Thus, through this simple automated process, the police analyst has potentially more information that may be pertinent to an OC and more specific information regarding the nature of the OC. Because the original OC Concept involved corroboration by a large number of sources, the analyst can gain some confidence that further information contained in sub-sets of the Tweets has credibility. Indeed, the analyst may now want to trace back to the original Tweets (or to the text of these Tweets) and, because they have been grouped together by FCA, it is simple task to provide this facility.

7 Implementation

The processes and components described above were implemented as a part of the European ePOOLICE Project [4]. The OC Taxonomy and entity extraction components developed by the authors (with assistance from people acknowledged below) were implemented in the system to provide data to be consumed by various analytic components, one of which was the FCA 'OC Threat Corroboration' component described above. The user interface to the system was developed by other colleagues at Sheffield Hallam University (who are also acknowledged below). The system allows a police analyst to select a region and type of OC to scan for and then acquire sources on the Internet (such as Tweets) that match those search criteria. Structured data is extracted automatically from the sources, as described above, allowing the user to carry out a variety of analytic tasks and display the results in an appropriate visualisation. For many of the analytics, including the FCA Threat Corroboration a map-based visualisation is used. Figure 8 shows the user interface with the FCA 'Corroborated Threats' option selected. There are a number of other options listed and these components were developed by other members of the ePOOLICE consortium. It is out of scope for this paper to describe them here, but for more information please visit the project website [4].

The map of the USA in Fig. 8 is displaying the San Diego and Atlanta OC Concepts from Fig. 6. Various icons are used by the system to indicate types of OC and the one here is for Human Trafficking. The analyst is able to select an OC Concept to display its information and to drill down to its sub-concepts. The Atlanta OC Concept has been selected and is thus displaying its associated information, including its attributes (*crime: humantrafficking* and *location: atlanta*) and its objects, listed as URLs. The 'drill-down' sub-concepts (from Fig. 7) are being displayed as icons below the main concept and Fig. 9 shows the additional information shown when one of these sub-concepts is selected - in this case the attribute *drug: amphetamine*. The 10 sources that contain a reference to amphetamine are listed and at this point the analyst may wish to look at some of these. Clicking on a URL will take the user to the original source.

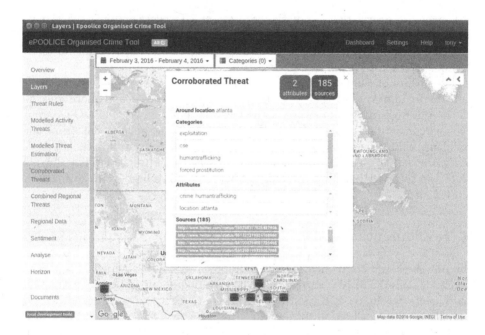

Fig. 8. ePOOLICE system showing San Diego and Atlanta OC Concepts and details of the Atlanta OC Concept

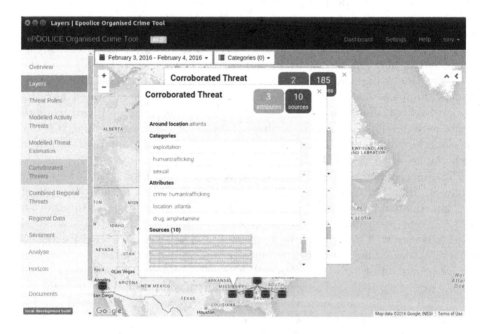

Fig. 9. ePOOLICE system showing Drill-Down information

7.1   Evaluation and Concluding Remarks

Although difficult to evaluation in an operational sense (we cannot, for example, act as the police in investigating organised crime) it is possible to say something about the quality of the results in terms of the accuracy of the weak signals. A sample of 20 inferred OC categories were inspected against the original text sources, with 16 out of 20 correctly identified from weak signals as being crime related. In the other four cases, the context within which the identified words or phrases were used clearly indicated that the source was not referring to OC. Although only a small sample, this was an encouraging level of false positives. Further evaluation is required, however, on larger samples to produce a statistically significant result.

A qualitative evaluation was provided by 24 end-users during a hands-on feedback session held in December 2015. Various law enforcement agencies from within the EU were represented, including those from a regional, national and international organisations. The challenges for the use of the system operationally centred on factors such as the need to refer to personal and sensitive information about the origins of the information extracted and corroborated using the system, and the complexity of the way in which the systems outputs were presented via the map-based interface. However, the utility of the system in providing a means to show trends in current, and alluding to emerging, forms of criminality in different geographic areas was overwhelmingly positive, especially through the utilisation of previously untapped data sources such as social media.

Acknowledgments. This project has received funding from European Union Seventh Framework Pro-gramme FP7/2007 - 2013 under grant agreement n FP7-SEC-2012-312651.

Legal and Ethical Disclaimer. No data that can or may be considered sensitive or personal has been handled as a result of the research undertaken. The authors do however acknowledge, despite being outside of the scope of the research present, that in practice the operational utility of such a system would be dependant on the use of data they may be considered personal and/or sensitive.

References

1. Andrews, S.: In-Close2, a high performance formal concept miner. In: Andrews, S., Polovina, S., Hill, R., Akhgar, B. (eds.) ICCS-ConceptStruct 2011. LNCS, vol. 6828, pp. 50–62. Springer, Heidelberg (2011)
2. Brewster, B., Ingle, T., Rankin, G.: Crawling open-source data for indicators of human trafficking. In: Proceedings of the 2014 IEEE/ACM 7th International Conference on Utility and Cloud Computing, pp. 714–719 (2014)
3. Chakraborty, G., Pagolu, M., Garla, S.: Text Mining and Analysis: Practical Methods, Examples, and Case Studies Using SAS. SAS Institute Inc., Cary (2014)
4. ePOOLICE. The epoolice project (2015)
5. Europol. Eu organised crime threat assessment: Octa 2011. file no. 2530–274. Technical report, Europol, O2 Analysis & Knowledge, The Hague (2011)

6. Ganter, B., Wille, R., Analysis, F.C.: Formal Concept Analysis: Mathematical Foundations. Springer-Verlag, New York (1998)
7. Twitter Inc., Twitter search api (2015)
8. Laczko, F., Gramegna, M.A.: Developing better indicators of human trafficking. Brown J. World Aff. **10**, 179 (2003)
9. Her Majesty's Inspectorate of Constabulary. The rules of engagement: A review of the disorders, August 2011
10. Omand, D., Bartlett, J., Miller, C.: Introducing social media intelligence (SOCMINT). Intell. Natl. Secur. **27**(6), 801–823 (2012)
11. Owoputi, O., O'Connor, B., Dyer, C., Gimpel, K., Schneider, N., Smith, N.A.: Improved part-of-speech tagging for online conversational text with word clusters. Association for Computational Linguistics (2013)
12. Perrin, A.: Social media usage: 2005–2015. Technical report, Pew Research Center (2015)
13. Ritter, A., Etzioni, O., Clark, S., et al.: Open domain event extraction from twitter. In: Proceedings of the 18th ACM SIGKDD international conference on Knowledge discovery and data mining, pp. 1104–1112. ACM (2012)
14. CISC Strategic Criminal Analytical Services: Strategic early warning for criminal intelligence. Technical report, Criminal Intelligence Service Canada (CISC) (2007)
15. United Nations. United nations convention against transnational organized crime and the protocols thereto
16. UNODC. Anti-human trafficking manual for criminal justice practitioners (2009)
17. UNODC. Global report on trafficking in persons (2009)
18. Williams, P., Godson, R.: Anticipating organized and transnational crime. Crime Law Soc. Change **37**(4), 311–355 (2002)

Distilling Conceptual Structures from Weblog Data Using Polyadic FCA

Sanda-Maria Dragoş[(✉)], Diana-Florina Haliţă, and Christian Săcărea

Department of Computer Science, Babeş-Bolyai University, Cluj-Napoca, Romania
sanda@cs.ubbcluj.ro, diana.halita@ubbcluj.ro, csacarea@math.ubbcluj.ro

Abstract. Formal Concept Analysis (FCA) is a prominent field of applied mathematics which is closely related to knowledge discovery, processing and representation. We consider the problem of distilling relevant conceptual structures from weblog data, more precisely, we investigate users' behavioral patterns in an web based educational platform by using n-adic FCA ($n = 3, n = 4$). We focus in our research on log data gathered from e-learning platforms. Such systems are particularly interesting, since user's behavioral patterns are closely related to their academic performance. We investigate user's behavior by using similarity measures of various visited page chains. We exemplify the methods we have developed on a locally developed e-learning platform called PULSE. Data gathered from weblogs have been preprocessed and conceptual landscapes of knowledge have been built using FCA. Triadic FCA (3FCA) is used to investigate correlations between similar page chains and the time granule when a certain pattern occurs. Finally, we employ tetradic FCA (4FCA) to compare web usage patterns wrt. temporal development and occurence. As far as we know, this is the first attempt to use 4FCA in web usage mining.

Keywords: Web usage mining · Behavioral patterns · Formal Concept Analysis · Similarity measures

1 Introduction

Weblogs are usually data strings comprising in a concise format all relevant information related to users activity on a certain website. The challenge to distill valuable knowledge is not minor. On the one hand, the scope of web usage mining is to gather relevant information related to how users are behaving on a certain web platform. On the other hand, weblogs have their own logic and are quite often raw data sets which contain in a rather hidden way the knowledge.

D.-F. Haliţă—This paper is a result of a doctoral research made possible by the financial support of the Sectoral Operational Programme for Human Resources Development 2007–2013, co-financed by the European Social Fund, under the project POSDRU/187/1.5/S/155383- "Quality, excellence, transnational mobility in doctoral research".

© Springer International Publishing Switzerland 2016
O. Haemmerlé et al. (Eds.): ICCS 2016, LNAI 9717, pp. 151–159, 2016.
DOI: 10.1007/978-3-319-40985-6_12

In this paper, we have focused on detecting repetitive browsing habits by using Formal Concept Analysis (FCA) [1]. Our purpose is to determine students with similar on-line access behavior; in that way we may find bundles of user with similar interests. Also, to identify common sub-behaviors of more complex behaviors, meaning common sub-chains of more complex visiting chains. The motivation of using n-adic FCA came from different situations demanding a fourth dimension for the underlying data structure. The browsing behavior was captured in so-called page chains (i.e., sequences of visited pages where the accessed page becomes the referrer for the next one).

In previous works, we have discovered different types of browsing patterns (relaxed, normal, intense - which give a quantitative description of how students are using the educational resources) [2] or attractors (more qualitative behavioral patterns to which users adhere while using the web-based educational system) [3]. The raw data is stored in weblogs. Conceptual scales are defined for every relevant attribute. These scales are then used to analyse the basic conceptual structures of weblogs. Then, we aggregate some of these scales into a triadic view, which reveals another insight of browsing patterns. A detailed discussion of various methods in behavioral pattern mining in web based educational systems is made in [4]. In this paper, we follow a similar approach but we discuss the use of polyadic FCA for mining *repetitive* behavioral patterns. First, we analyze similarity of visited page chains and then we discuss how triadic and tetradic concepts express relevant knowledge clustering of repetitive behavioral patterns. Triadic FCA is used to analyze these patterns for single users, while tetradic FCA can be used to study repetitiveness for pairs of users w.r.t. some meta-classes.

While recent contributions are focusing on behavioral pattern analysis in well-established learning platforms [5], we have used PULSE [6]. This portal, which is personalized for every user that enters it, was mainly designed to be used for presenting theoretical support for the studied subjects and automatically setting assignments and recording evaluations for individual work and tests. The system was progressively built and enhanced according to its users needs. This gave us the necessary access to improve not only the educational content on PULSE but also to its design, since we wanted to have an informed learning management system that continually educates itself about the requirements of its users as a result of the feedback offered by various pattern mining tools and thus to evaluate the effectiveness of PULSE.

Due to the strength of its knowledge discovery capabilities and the subsequent efficient algorithms FCA seems to be particularly suitable for analysing educational sites. For instance, papers [7,8] are devoted to the topic of improving discussion forums, while our own previous contributions are focusing on the user/student behavior [2,3].

The work presented in this paper is intended as a proof of concept for applying higher-adic FCA on web data. Moreover, current FCA instruments have scalability limitations and therefore we had to reduce the scale of our data. We have come to reach the limits of these instruments during our tests. Our initial goal (started in 2010) was to improve PULSE and we used since then many methods to do that (including statistical tools), but most of them are FCA based. However, our current goal is to observe and understand student behavior and the interaction between them while using PULSE with the expressiveness power and capabilities of FCA.

2 Building the Contexts

We considered each session as a chronologically ordered sequence of page accesses, which we called chains. On these chains, we used different similarity measures in order to find similar patterns of usage behavior. Our purpose is to: determine students with similar on-line access behavior in order to find bundles of users with similar interests; identify students (that we call "trendsetters") who initiate behaviors than influence the behavior of other users; group pages with similar access or usage - in that way the designer of the e-learning platform may group information according to users' needs, and identify common sub-behaviors of more complex behaviors.

For each student we have constructed chains formed by pages visited during a session and we have associated to them the corresponding week as an aggregation of the visit timestamp. Then we have compared these chains in order to determine students' repetitive behavior. We also compared chains of different users to identify the influence one user may have over others and to eventually determine possible "trend-setters" as we have defined them in [3]. For comparing these chains, we have used the Cosine similarity measure [9]. The objects are chains that have a Cosine similarity of at least 80 %. Attributes, conditions, states, etc. are then chosen appropriately, as described below.

3 The Triadic Approach

In this section, we describe how tricontexts are build and how triconcept sets can be visualized using a local, dyadic perspective with the aim to determine some single-user repetitive behavior. These users have been selected from those who adhere to previously studied educational attractors. For this, we need to aggregate the pages that appeared in similar chains as they have been previously determined. We focussed on 10 meta-access classes: information about the lecture and the laboratories (I), all lab assignments (LA), information about the practical examination (PE), all laboratory examples (LE), theoretical support for all laboratories (LT), information about the written examination (WE), overview information about the test papers given during lectures (LP), details on all lecture test papers (LPs), overview information about the lectures (L), slides and notes for all lecture (Ls). We determined for all users the chains that have a similarity of at least 80 %. That gave us pairs of similar chains that occurred within certain weeks. Next, we substituted all chains with a code showing which out of these 10 classes are present in that chain.

Triadic FCA (3FCA) is a natural extension of the classical dyadic case, where objects are related to attributes and conditions by a ternary incidence relation.

Unfortunately, the classical representation of a triconcept set as a trilattice has several drawbacks, the major one being the lack of graphical expressive power. Paper [10] proposes a new navigation paradigm in triconcepts sets by locally projecting along one dimension which is called *perspective*. This reduces the task of visualization of the entire triconcepts set to a local navigation problem in concept lattices obtained by projecting the tricontext along a certain *perspective*.

Selecting a student named K, generates a list of 20 non-trivial triconcepts showing different classes of repetitive behavior of K. A brief listing of these triconcepts does not give an insight of the connections within these triadic conceptual structures. In order to visualise them, we start with some triconcept, say $([L - LPs, L - LPs - WE], [w1 = 15, w1 = 16], [w2 = 16])$ and project along the third perspective, i.e., the conditions set. We obtain the diagram represented in Fig. 1(a) which shows the conceptual structure obtained by focussing on week 16, i.e., the end of the semester and projecting the triadic data set along this set. Here, we need to emphasize that all nodes in Fig. 1(a), (b) correspond to some triconcepts generated for user K.

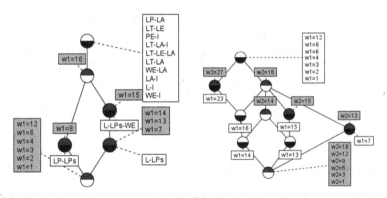

(a) Navigation along perspective 3, (b) Navigation along perspective 1, condition $w2=16$ chain $[L\text{-}LPs]$

Fig. 1. Local navigation for user K

Here it can be more easily seen that for user K, none of the chains that reoccurred in week 16 have not occurred in weeks 1–4, 6 or 12. We also observe that chains containing LP-LPs that reoccurred in week 16 first occurred only in week 8. But, chains containing L-LPs that reoccurred in week 16 first occurred in weeks 7, 13 or 14. We also can see that L-LPs is a subchain of L-LPs-WE and therefore they are positioned in a subordinative relation in the diagram. All chains within the top most node never reoccurred in week 16.

Navigation can continue now, by focussing on another concept. Choosing the rightmost down node, i.e. concept with extent $[L - LPs]$, we project now along perspective 1, i.e., the object set. This switches to the following diagram represented in Fig. 1(b). From this concept lattice, we can easily observe the pairs of occurrence and reoccurrence weeks for user K. One may go up to collect all the reoccurrence weeks and down to collect all the occurrence weeks. For instance for the first occurrence in week 14, the chain L-LPs reoccurred in weeks 14, 16 and 27. Navigation can continue by choosing any other node and one of the three perspectives.

4 An Extension: 4FCA

Polyadic FCA has been introduced in [11] as a natural extension of FCA to n-adic data sets. Similar to the dyadic case, we can compute n-concepts from n-contexts.

Definition 1. *An n-context is an $(n+1)$-tuple $\mathbb{K} := (K_1, ..., K_n, Y)$ with $K_1, ..., K_n$ being sets and $Y \subseteq K_1 \times ... \times K_n$ an n-ary incidence relation. An n-concept of an n-context $\mathbb{K}$ is an n-tuple $(A_1, ..., A_n)$ satisfying $A_1 \times ... \times A_n \subseteq Y$ and for every $n-tuple$ $(C_1, ..., C_n)$ with $C_i \subseteq A_i, \forall i \in \{1, ..., n\}$ the following relation is satisfied: $C_1 \times ... \times C_n \subseteq Y \Rightarrow C_i = A_i, \forall i \in \{1, ..., n\}$.*

For $n = 4$, we consider tetradic data sets, i.e., objects, attributes, conditions and states related by an 4-adic incidence relation. In order to compute 4-concepts, we generalize the well-known Trias algorithm for computing triconcepts. For each state, we consider the corresponding tricontext and compute first all triconcepts of that "slice". Then we complete the 3D cuboids to the fourth dimension, by imposing the maximality condition of Definition 1.

```
 1  Let O,A,C be three sets, representing objects, attributes and
        conditions which are common for all the tricontexts.
 2  Let K_1 := (K_11, K_12, K_13, Y_1), ..., K_n := (K_n1, K_n2, K_n3, Y_n) be n tricontexts with
        K_i1 ⊆ O, K_i2 ⊆ A, K_i3 ⊆ C and Y_i ⊆ K_i1 × K_i2 × K_i3 the ternary relation
        between them, ∀i ∈ {1, ..., n}.
 3  Θ = ∅;
 4  ∀ K_i, i ∈ {1, ..., n}
 5    A_4 = state(K_i);
 6    for all (A_1, A_2, A_3) ⊆ Y_i do add(Θ, (A_1, A_2, A_3, A_4));
 7  ∀ (θ_i, θ_j) ⊆ Θ × Θ
 8    if (θ_i(A_4) ⊆ θ_j(A_4) & ∃l, m ∈ {1, 2, 3}, l ≠ m so that θ_i(A_l) ∩ θ_j(A_l) ≠ ∅ &
          θ_i(A_m) ∩ θ_j(A_m) ≠ ∅) then θ_i = θ_i ∪ θ_j;
 9  ∀ (θ_i, θ_j) ⊆ Θ × Θ
10    if (θ_i(A_4) == θ_j(A_4) & ∃l, m ∈ {1, 2, 3}, l ≠ m so that θ_i(A_l) == θ_j(A_l) &
          θ_i(A_m) == θ_j(A_m)) then θ_i = θ_i ∪ θ_j; & delete(θ_j);
11  ∀ (θ_i, θ_j) ⊆ Θ × Θ
12    if (θ_i == θ_j) then delete(θ_i);
13  ∀ (θ_i, θ_j) ⊆ Θ × Θ
14    if (θ_i(A_4) == θ_j(A_4) & ∃l, m, n ∈ {1, 2, 3}, l ≠ m ≠ n ≠ l so that θ_i(A_l) == θ_j(A_l)
          & θ_i(A_m) == θ_j(A_m) & θ_i(A_n) ⊆ θ_j(A_n)) then delete(θ_i);
15  ∀ θ_i ⊆ Θ & ∀ K_j, j ∈ {1, ..., n} & ∀ (A_1, A_2, A_3) ⊆ Y_j
16    if (A_1 == θ(A_1) & A_2 == θ(A_2) & A_3 == θ(A_3) & state(K_j) == θ(A_4)) then
          save(i)=true;
17  ∀ θ_i ⊆ Θ
18    if (!save(i)) then delete(θ_i);
19  Θ is the list of all tetraconcepts derived from all the triconcepts.
```

Listing 1.1. The pseudocode of the algorithm used for generating tetraconcepts

4.1   The Tetradic Approach

To exemplify this approach on a small example, we considered a group of students from the same specialization and from the same year. We namecoded them from A to X. For each student we compare their e-learning platform browsing habits, with all the others. We are interested in evaluating repetitive behavioral patterns, hence we focus, as previously stated on chains that have Cosine similarity of at least 80 %. By this comparison, we obtain pairs of chains, corresponding pairs of users, as well as the time granule, i.e., the weeks in which those chains were generated. For each user X, we obtain by this a 4-context and a list of corresponding 4-concepts:

- The *objects* set contains all users having similar chains with X, including X;
- The *attributes* set contains the weeks in which the chain occurred as a behavior of X;
- The *condition* set contains the weeks in which the same behavior was detected at users from the objects set;
- The *states* are the list of actual *behaviors* as encoded chains.

By this tetradic approach, we can detect up to 15 distinct chains (i.e., behaviors) for the active users. Each such chain contains pairs of two or three meta-access classes out of those 10 we mentioned at the beginning of Sect. 3 and every such chain may be related to other "subchains" which contain even more meta-access classes.

```
1  ([u2=H,K,O,T], [w1=2], [w2=2], [LA–I])
2  ([u2=E], [w1=1,w1=2,w1=3], [w2=6], [LA–I])
3  ([u2=K], [w1=2], [w2=2], [LA–I,WE–LA–I])
4  ([u2=B,H,T], [w1=3], [w2=2], [LT–LA–I])
5  ([u2=H], [w1=3], [w2=2], [LT–LA–I,LT–LA])
6  ([u2=K], [w1=3], [w2=3], [LT–LA–I,L–LA–I,LT–LA,L–LA,L–WE–LA])
7  ([u2=O], [w1=3], [w2=2], [L–LA–I])
8  ([u2=B], [w1=9], [w2=7], [L–PE–LA–I])
9  ([u2=K], [w1=16], [w2=16], [LP–LPs–PE–LA–I,LP–LPs,L–LPs,L–WE,Ls–L–LP
       ])
```

Listing 1.2. Tetradic analysis for user K for the chain *"LA-I"*

As depicted in Listing 1.2, in the second week, there are 4 students that used the educational content of the platform in a similar way. They have at least one 80 % similar chain with a chain of K, i.e., they prove to have a similar behavior. In other cases, user K has the same behavioral pattern over several weeks (i.e., the intent set of the 4-concept is no longer a singleton) and/or a repetitive behavior throughout different weeks for the other students (i.e., the modi set is not a singleton). For instance, from the second tetraconcept from Listing 1.2 we observe that user E has a similar visiting chain (i.e., behavior) in week 6 as K had in weeks 1, 2 and 3. In other cases, the status subset is not a singleton, i.e., contains more chain-encoded behaviors. These behaviors can be related (i.e., including the initial behavior) or not. The related behaviors in this case appear in all last seven 4-concepts of Listing 1.2. For instance, the tetradic concept 4 refers to users B, H, and T which visited Lecture Theory, Lab Assignment and Info pages in week 2 with at least 80 % similarity to the behavior that K exhibits in week 3. What is also interesting here is to note that user K exhibits in week 3 behaviors that repeated within the same week out of 5 different classes of interest, as it can be discovered from analyzing tetraconcept 6. We also have to mention that some 4-concepts may contain correlations between 4-adic data that are not related with the initial class (e.g. 4-concept 9). In week 16, user K shows a repetitive behavior which maps into 5 chains, out of which the last 4 are unrelated to the starting chain "LA-I".

We selected two most active users (K and H), starting by assessing their results from a quantitative perspective. For each of these two students we have determined the bundle of users that have the same behaviour with them. For instance, for chain "Lecture, Lecture Paper", these bundles are shown in Listing 1.3.

```
1    For  student  K
2       users  with  similar  behavior  =  G,  H,  K
3       weeks  of  behavioural  occurrence  for  K  =  7,  13
4       weeks  of  behavioural  occurrence  for  users  with  similar  behavior
           with  K  =  5,  6,  7,  11,  14
5
6    For  student  H
7       users  with  similar  behavior  =  B,  D,  F,  G,  H,  I,  W,  K,  L,  M,  O,  Q
           ,  T,  X
8       weeks  of  behavioural  occurrence  for  H  =  4,  5,  6,  7,  9
9       weeks  of  behavioural  occurrence  for  users  with  similar  behavior
           with  H  =  3,  4,  5,  6,  7,  8,  9,  11,  12,  14,  15,  17,  20
```

Listing 1.3. Stats for student K on chain *"L-LP"*

We can discover here from a common bundle formed by students G, H and K. Another aspect is that this type of behavior appeared for the first time in week 3 and neither of students H or K generated it. We determined that student M was first to have this behavior (i.e., in week 3).

We also determined for both H and K the other students that have similar behaviors that fit into different classes. The bundle of students that have repeated the behavior of student K in at least 9 distinct chains are: G, H and V. For student H we have found students: K, G, D and I. We can conclude that overall students G, H and K have very similar behaviors. However, the details within the FCA context show how different students H and K are. Preserving such important details is the main strength of FCA.

5   Conclusions and Future Work

In this paper, we have discussed how higher-adic FCA can be used to detect a various palette of repetitive behavioral patterns and its usability in Web Usage Mining. We have exemplified here the manner in which one may investigate 4-adic FCA results by using both quantitative and qualitative measurements. We have switched between quantitative methods for a bird-eyes view and then returned to qualitative technics (such as FCA) to see the actual facts. There are also some limitations of the approach, mainly due to insufficient developed navigation and exploration features of higher-adic FCA. This contribution is a preliminary work and a starting point for a research project of distilling conceptual structures using higher-adic FCA from raw weblog data. We have managed to determine similar behavior that fit into classes of our interest. We also discovered which are the actual meta-classes that users use. Based on an analysis of users showing a *common behavior*, we determined bundle of users with similar behavior. We are very close to determine trend-setters, but we need to have another perspective on our data. For that we have to put together all our data and analyse it on a 5-adic perspective.

References

1. Wille, R.: Conceptual landscapes of knowledge a pragmatic paradigm for knowledge processing. In: Gaul, W., Locarek-Junge, H. (eds.) Proceedings of the 22nd Annual GfKl Conference on Classification in the Information Age, Dresden, 4–6 March 1998, Berlin, Heidelberg, pp. 344–356. Springer, Heidelberg (1999). http://dx.org/10.1007/978-3-642-60187-3_36
2. Dragoş, S., Haliţă, D., Săcărea, C., Troancă, D.: Applying triadic FCA in studying web usage behaviors. In: Buchmann, R., Kifor, C.V., Yu, J. (eds.) KSEM 2014. LNCS, vol. 8793, pp. 73–80. Springer, Heidelberg (2014)
3. Dragos, S., Halita, D., Sacarea, C.: Attractors in web based educational systems a conceptual knowledge processing grounded approach. In: Zhang, S., Wirsing, M., Zhang, Z. (eds.) KSEM 2015. LNCS, vol. 9403, pp. 190–195. Springer, Heidelberg (2015). doi:10.1007/978-3-319-25159-2_18
4. S. Dragos, D. Halita, C. Sacarea.: Behavioral pattern mining in web based educational systems. In: Rozic, N., Begusic, D., Saric, M., Solic, P. (eds.) 23rd International Conference on Software, Telecommunications, Computer Networks, SoftCOM 2015, Split, Croatia, 16–18 September 2015, pp. 215–219. IEEE (2015). http://dx.org/10.1109/SOFTCOM.2015.7314076
5. Romero, C., Espejo, P.G., Zafra, A., Romero, J.R., Ventura, S.: Web usage mining for predicting final marks of students that use moodle courses. Comput. Appl. Eng. Educ. **21**(1), 135–146 (2013)
6. S. Dragos.: PULSE extended. In: The Fourth International Conference on Internet, Web Applications, Services, Venice/Mestre, Italy, pp. 510–515. IEEE Computer Society, May 2009
7. Cerulo, L., Distante, D.: Topic-driven semi-automatic reorganization of online discussion forums: a case study in an e-learning context. In: Global Engineering Education Conference (EDUCON), pp. 303–310. IEEE (2013)
8. Distante, D., Fernandez, A., Cerulo, L., Visaggio, A.: Enhancing online discussion forums with topic-driven content search and assisted posting. In: Fred, A., Dietz, J.L.G., Aviero, D., Liu, K., Filipe, J. (eds.) Knowledge Discovery, Knowledge Engineering and Knowledge Management, vol. 553, pp. 161–180. Springer. New York (2014)
9. Gan, G., Ma, C., Wu, J.: Data Clustering: Theory, Algorithms, and Applications, vol. 20. Siam, Philadelphia (2007)

10. Rudolph, S., Săcărea, C., Troancă, D.: Towards a navigation paradigm for triadic concepts. In: Baixeries, J., Sacarea, C., Ojeda-Aciego, M. (eds.) ICFCA 2015. LNCS, vol. 9113, pp. 252–267. Springer, Heidelberg (2015)
11. Voutsadakis, G.: Polyadic concept analysis. Order **19**(3), 295–304 (2002). http://dx.org/10.1023/A:1021252203599

Ontologies and Linked Data

$[MS]^2O$ – A Multi-scale and Multi-step Ontology for Transformation Processes: Application to Micro-Organisms

Juliette Dibie[1], Stéphane Dervaux[1], Estelle Doriot[1], Liliana Ibanescu[1(✉)], and Caroline Pénicaud[2]

[1] UMR MIA-Paris, AgroParisTech, INRA, Université Paris-Saclay,
75005 Paris, France
{juliette.dibie,stephane.dervaux,liliana.ibanescu}@agroparistech.fr
[2] UMR GMPA, AgroParisTech, INRA, Université Paris-Saclay, 78850
Thiverval-Grignon, France
caroline.penicaud@grignon.inra.fr

Abstract. This paper focuses on the knowledge representation for an interdisciplinary project concerning transformation processes in food science. The use case concerns the production of stabilized micro-organisms performed at INRA (French National Institute for Agricultural Research). Experimental observations are available for some inputs of the production processes, at different steps and at a certain scale. Available data sets are described using different vocabularies and are stored in different formats. Therefore there is a need to define an ontology, called $[MS]^2O$, as a common and standardized vocabulary. Users' requirements were defined through competency questions and the ontology was validated against these competency questions. $[MS]^2O$ ontology aims to play a key role as the representation layer of the querying and simulation systems of the project. This leads to the possibility of comparing different production scenarios and suggesting improvements.

Keywords: Domain ontology building · Multi-step and multi-scale ontology · Transformation processes

1 Introduction

There is a challenging need for food companies to design and to control the transformation of raw materials into final products, ensuring their quality while applying appropriate technologies for minimal economic, environmental and social costs. For that, they have to better understand the food production system in order to adopt an eco design approach by considering concomitantly product quality, production process parameters, and its environmental impacts. Such an eco design approach supposes to take advantage of all available data and experts' knowledge for performing the analysis of the production system. However, data have been collected for different purposes, in different sub-domains,

O. Haemmerlé et al. (Eds.): ICCS 2016, LNAI 9717, pp. 163–176, 2016.
DOI: 10.1007/978-3-319-40985-6_13

at different scales. Data have also been encoded in various formats using heterogeneous vocabularies and are processed in different information systems. Moreover, expert's knowledge is often implicit and difficult to acquire. There is therefore a need to uniformly model and store available data and experts' knowledge in order to compare different production scenarios and to perform a cause and effect analysis.

Ontologies are nowadays used as a common and standardized vocabulary for representing concepts and relations from a particular field (e.g. life-science, geography). An ontology is designed to represent the knowledge from a domain in terms of concepts, relations between these concepts and instances of these concepts [8]. Building networks of interconnected ontologies [5] and publishing ontologies on the Linked Open Data (LOD) cloud[1] should facilitate data integration and data sharing, such as giving access to data from specific disciplines or data produced within specific geographic regions [2]. However, domain ontologies are built for a specific task and it is not easy to reuse them for a different one. Data available at different scales and the challenge to take into account the environmental impact of the food production process cause the knowledge management in food production more complex.

This paper focuses on building an ontology, called $[MS]^2O$, for transformation processes and more particularly the production of stabilized microorganisms, performed at INRA (French National Institute for Agricultural Research) in an inter disciplinary project called LIONES.

A process may be represented as an industrial process [7], a business process [12] or a food transformation process [10]. However, none of these representations was suited to represent a food transformation process described by a set of experimental observations available at different scales. Different Ontology Design Patterns (ODP)[2] exist as for example the ODP for material transformation [13] or the ODP for life cycle assessment data [9]. As discussed in [3], ODPs bring a promise of compositional modeling and true ontology reuse, but many barriers to their adoption still remain.

Building the ontology from scratch, one of the scenarios of the NeON methodology [5] was used for the construction of $[MS]^2O$. Users' requirements were defined through competency questions and the ontology was validated against these competency questions.

$[MS]^2O$ core component is implemented in OWL[3] and all the available data were structured in files using the $[MS]^2O$ vocabulary. The domain component of $[MS]^2O$ is under development.

The paper is organized according to the NeOn methodology. In Sect. 2 the ontology specification is presented, then the ontology conceptualization is detailed in Sect. 3. In Sect. 4, we present the ontology implementation and user evaluation. Finally, we conclude in Sect. 5 and present our further work.

---

[1] http://linkeddata.org.
[2] http://ontologydesignpatterns.org.
[3] https://www.w3.org/2001/sw/wiki/OWL.

2   Ontology Specification

This section briefly presents the LIONES Project, then the competency questions expressing users' requirements and finally our ontology specification.

2.1   LIONES Project

LIONES Project involves domain experts and computer scientists researchers from INRA, the French National Institute for Agricultural Research. It addresses the issue of modeling semantic **LI**nks between **ON**tological multi-scal**ES** objects involved in transformation process. LIONES Project is applied to the production of stabilized micro-organisms. Micro-organisms are biological agents which present a large scale of applications in food (e.g. yoghurts, cheese, wine, beer) and non-food (e.g. probiotics, microbial production of chemical molecules, bio-fuels) domains. The need for concentrated micro-organisms (called starters) to be stabilized and in ready-to-use form is increasing. Their production process [1] is composed of a set of steps or unit operations (e.g. fermentation, cooling, concentration, formulation) that transforms inputs into outputs. The inputs of the system are the raw materials (e.g. micro-organisms, sugar), energy and water. The outputs of the system are the final product, some co-products, energy and effluents. Figure 1 presents the production system for stabilized micro-organism from the domain experts point of view. Two distinct axes were identified: the multi-step axis composed of the different steps of the production process and the multi-scale axis which represents the different scales of the studied product. Moreover, the multi-criteria analysis of the production system should help the experts to guarantee the product quality during the transformation process while reducing economical costs and environmental impact.

The modeling of semantic links in LIONES Project first requires to qualify and represent the multi-scales objects involved in the transformation process. This paper focus on the knowledge representation task of the LIONES Project: the building of $[MS]^2O$, a Multi-Scale and Multi-Step Ontology for modeling the production of stabilized micro-organisms.

The available data sets in LIONES Project concerns different steps of the production process at different scales, from the microbial cell components to the target functionality at the population level. Data sets are heterogeneous and sparse and come from many different sources such as databases, EXCEL files, text files, sensor outputs, scientific publications [6,11,14], Master reports, laboratory technical reports. They are gathered for many different purposes by different experts with their own experimental itineraries, vocabularies and technical materiel and methods. Since the experts work on the same domain, reaching a consensus about a common vocabulary was quite easy. Nevertheless (i) extracting and expressing implicit experts' knowledge, (ii) understanding and structuring the whole transformation process with all the involved entities, their properties and interactions using the available data and documents and the implicit experts' knowledge and (iii) identifying the users requirements for data retrieval

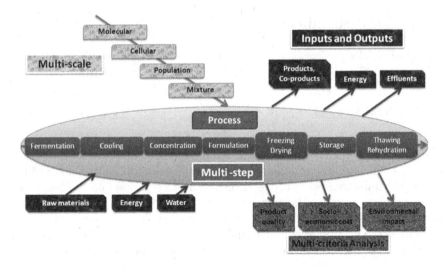

Fig. 1. The production system for stabilized micro-organism

and reasoning were less easier and required extensive conversations with domain experts to model the ontology.

In the next sub-section, we present the users' requirements through competency questions (CQ), which are natural language sentences that express a pattern for a type of questions domain experts expect the ontology to answer.

2.2   Competency Questions

Domain experts involved in LIONES Project have the following criteria for the product process analysis: the product quality, the environmental impact, and the economical cost. We helped the domain experts to express their users' requirements through a set of competency questions, as for example:

– Competency questions about product quality analysis
   CQ_1 Find all the production conditions of stabilized *Lactobacillus delbrueckii* subspecies *bulgaricus* CFL1 that allowed obtaining a specific acidifying activity lower than 20 min/log (CFU/mL).
   CQ_2 Find all the datasets about cultivability losses and membrane integrity of *Yarrowia lipolytica* during drying.
   CQ_3 Find the Yeast having the most cultivability losses during drying.
   CQ_4 Find which process step was the most damaging for the micro-organisms viability.
   CQ_5 Find all the datasets about drying resistance of *Yarrowia lipolytica* during the fermentation step.
– Competency questions about environmental impact analysis
   CQ_6 Find all the process steps that are involved in global warming.
   CQ_7 Find the energy consumptions of freeze-drying measured in France during the year 2014.

– Competency questions about economical cost analysis
     CQ_8 Find all the freeze-drying conditions that allowed producing more than
          10t/month of stabilized bacteria.

The next sub-section presents the ontology specification aiming to answer to
the domain experts' competency questions.

2.3   Ontology Specification

Using the Fig. 1, the available data sets and the related documents, we established, in close collaboration with domain experts, a new representation of the
production system as given in Fig. 2.

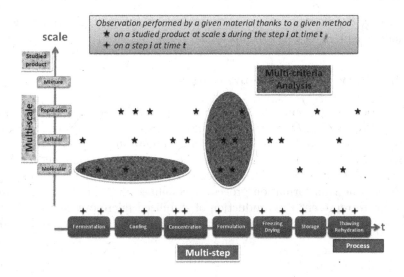

Fig. 2. Global schema of the production system

In this new schema, the two axes of Fig. 1 remain: the multi-step axis where
the process is described by its different steps and the multi-scale axis where the
studied products can be considered at different scales (i.e. as a whole inside a
mixture, at the population scale, the cellular or molecular ones). Moreover, we
give the following definitions:

D1 **a transformation process is composed of steps, non necessarily in
    sequence** (e.g. two steps can occur simultaneously), **which transforms
    inputs into several different outputs;**
D2 **the representation of a transformation process relies on experimen-
    tal observations;**

D3 **experimental observations are performed with a given material according to a given method either on a mixture (i.e. a set of combined products) at a certain scale during a certain step** (e.g. observations about the membrane integrity of *Yarrowia lipolytica* performed at its cellular scale during the drying step in CQ_2) **or on a step** (e.g. observations about energy consumption by the freeze-drying step in CQ_7).

D4 **the inputs and outputs of the system** (e.g. raw materials, water, co-products and effluents) of Fig. 1 **are components of the mixture and can be evaluated through observations on it at different steps and different scales.**

D5 **The product quality, the economic cost and the environmental impact** of Fig. 1 **can be deduced from attributes associated with the studied products, the steps of the process and the materials.**

These five definitions correspond to our understanding of the studied domain and guide our $[MS]^2O$ conceptualization.

3   Ontology Conceptualization

In this section, we present our Multi Scale and Multi Step Ontology for transformation Processes, $[MS]^2O$. Our ontology is designed from scratch by an iterative process in a modular way in order to be able to import and reuse existing ontologies, which will be the next step of our work, as recommended by the NeOn methodology. $[MS]^2O$ is composed of two components: a core component to represent the transformation processes as defined in D1 to D5 and a domain component to represent the production of stabilized micro-organisms. The core component of $[MS]^2O$ is composed of the five following modules (see Fig. 3):

- a Process module
- a Studied object module
- an Attribute module
- an Observation module
- a Material and Method module

In the next sub sections, we detail its main entities presented in Fig. 4, with examples on the production of stabilized micro-organisms. The Material and Method module is not presented in this paper since a complementary expertise is necessary to develop it. Even if its existence seems obvious for the domain experts, this module was not in the initial user requirements. The definition of the competency questions for this module is in progress.

3.1   The Process Module

According to definition D1, a transformation **process** is composed of steps, non necessarily in sequence (e.g. two steps can occur simultaneously) which transforms inputs into several different outputs. A **process**, which corresponds to

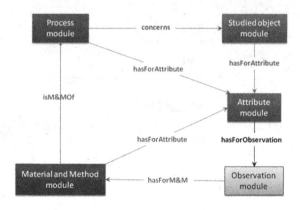

Fig. 3. The five modules of the $[MS]^2O$ core component

one experimentation for the domain experts, is composed of several itineraries (see `hasForItinerary` property in Fig. 4). An `itinerary` is composed of a set of steps (see `hasForStep` property in Fig. 4) characterized by some experimental conditions linked by time relations (see `TimeRelation` property on `step` concept in Fig. 4).

Example 1. Let us consider a process on the production and stabilization of the yeast Yarrowia lipolytica *performed with two distinct itineraries. The first itinerary is composed of the steps: fermentation, concentration, drying into stove during 75 min and storage. The second one is composed of: fermentation, concentration, drying into fluidized bed during 90 min and storage.*

A `step` may be composed of substeps (see `hasForSubstep` property in Fig. 4), which are considered as steps. A `step` is characterized by its material and method (see `hasForM&M` property in Fig. 4). Figure 5 gives an example of the possible substeps of the fermentation step in different cases.

Each `itinerary` and `step` is characterized by the time when it starts (see `hasForTimeProperty` property in Fig. 4).

A step can be compared with the perdurant object *Phenomenon* in DOLCE[4] and with the concept *Process* in SUMO[5].

3.2 The Studied Object Module

A studied object may be a **product** or a **mixture**. The studied product is the one on which is applied the transformation process (see `hasForStudiedObject` property between **process** and **product** in Fig. 4).

Example 2. The studied product of the process presented in Example 1 is the yeast: Yarrowia lipolytica.

---

[4] http://www.loa.istc.cnr.it/old/DOLCE.html.
[5] http://www.adampease.org/OP/.

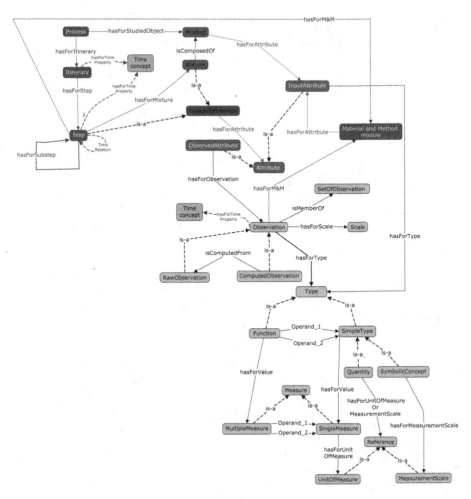

Fig. 4. Detail of the modules Process, Studied object, Attribute and Observation of the $[MS]^2O$ core component

A `mixture` is a composition of several products (see `isComposedOf` property in Fig. 4), but may be composed of only one product. A `mixture` is characterized by its composition (e.g. raw materials, water). It contains in particular the studied product. The different `steps` and sub-steps of an itinerary are applied on `mixtures` (see `hasForMixture` property in Fig. 4). Let us notice that a `step` may be applied on several `mixtures`.

A mixture can be compared with the DOLCE concept *Non-Agentive-Physical-Object*.

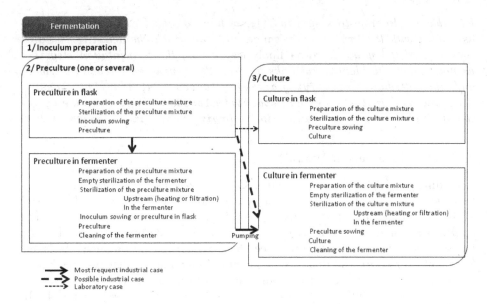

Fig. 5. The possible substeps of the fermentation step

3.3 The Attribute Module

The features of interest of our domain are the steps and the studied mixtures. They are characterized by `attributes` which may have input values or may be observed (see `hasForAttribute` property between `FeatureOfInterest` and `Attribute` in Fig. 4). An attribute can therefore be either an `input attribute`, it is then associated with input values (see Sect. 3.4), or an `observed attribute`, it is then associated with observations (see `hasForObservation` property between `ObservedAttribute` and `Observation` in Fig. 4). Let us notice that the material and method are also characterized by input attributes (see `hasForAttribute` property between `Material and Method module` and `Attribute` in Fig. 4).

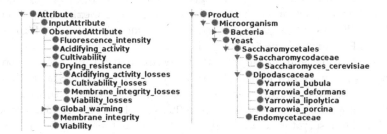

Fig. 6. An excerpt of the attribute hierarchy and the yeast hierarchy in the $[MS]^2O$ domain component applied to the production of stabilized micro-organisms

Example 3.   In order to answer to CQ_5, we have to consider all the observations associated with the Drying_resistance *attribute which characterizes mixtures composed of the yeast* Yarrowia lipolytica, *during the fermentation step. Let us notice that the hierarchical links between the* Drying_resistance *attribute and the attributes* Acidifying_activity_losses, Cultivability_losses, Membrane_integrity_losses *and* Viability_losses *in Fig. 6 eases the querying of all observations associated with the* Drying_resistance *attribute.*

3.4   The Observation Module

According to definitions D2 and D3, observations are very important for the transformation process representation. They correspond to the experimental results. Each observation is performed by a material (i.e. a sensor) using a given method (see hasForM&M property in Fig. 4), on a step or on a mixture (see hasForAttribute property between FeatureOfInterest and ObservedAttribute, then hasForObservation property between ObservedAttribute and Observation in Fig. 4) at a given scale (e.g. molecular, cellular, population) (see hasForScale property in Fig. 4). It is important to stress on, as states in definition D3, that the multi scale property of $[MS]^2O$ is represented at the observation level. As a matter of fact, in our domain, what is important is to know at which scale a given experimental measure (i.e. an observation) is made, not to register the part-of links between different parts/scales of a product. Currently, the observations are made independently at different steps and different scales of the studied mixtures.

Example 4.   The observations associated with the Cultivability_losses *attribute are performed on mixtures at their population scale. In order to answer to CQ_3, we have to consider all the observations, associated with the* Cultivability_losses *attribute, which are performed on mixtures composed of a yeast during the drying step (see the excerpt of the yeast hierarchy in Fig. 6).*

An observation can be a raw observation or a computed observation. A computed observation is computed from raw observations (see isComputed From property in Fig. 4). The observations are regrouped in a set of observations, which allows one to register that they were made together and represent the same set of experimental measures (see isMemberOf property in Fig. 4). An observation is characterized by the time when it is performed (see hasForTimeProperty property in Fig. 4).

An observation and an input attribute are defined by their type (see hasForType property in Fig. 4). This type can be either (i) a SimpleType, i.e. a quantity (e.g. temperature) or a symbolic concept (e.g. texture), or (ii) a function composed of two operands (see Operand_1 and Operand_2 properties in Fig. 4). The observation or the input attribute has then for value in the first case a single measure associated with a unit of measure, and, in the second case a multiple measure.

Example 5. The observations on the Membrane_integrity *attribute, involved in* CQ_2, *are computed from the raw observations associated with the* Fluorescence_ intensity *attribute. Each raw observation on the* Fluorescence_intensity *attribute has for type a function between the quantity* Fluorescence_intensity_ quantity *and the quantity* NumberOfCells. *Each computed observation on the* Membrane_integrity *attribute has a simple type, the boolean symbolic concept. It has for value a boolean value, with the unit of measure 'no_unit'.*

4   The Ontology Implementation and Validation

The current version of $[MS]^2O$ core component implemented in OWL is available at http://lovinra.inra.fr/2015/12/16/multi-scale-multi-step-ontology/.

Time ontology[6] recommended by the W3C[7] is used to represent the temporal concepts. For the measure part we reuse the modeling of the measures with their units from the @Web platform[8] [4] which was inspired from OM[9].

Available data concerning two production processes with two itineraries each were structured using the $[MS]^2O$ vocabulary and are stored into 162 EXCEL files. For each production process,

- 1 file describes the process,
- 66 files contain experimental observations, where the biggest experimental observation contains 297 results' raws,
- 4 files contain the mixture composition,
- 9 files describe the steps, and
- 1 file contains the environmental impact information.

Example 6. Figure 7 gives a screen-shot of an excerpt of the EXCEL file which describes a process.

These EXCEL files allow us to build the $[MS]^2O$ domain component and to validate that all available data in the use case can be represented as instances of $[MS]^2O$. To validate our conceptualization, we have checked that the SPARQL queries deduced from the competency questions presented in Sect. 2.2 have answers in a subset of OWL instances built from the EXCEL files. A knowledge base built from EXCEL files is currently in progress.

Example 7.   The SPARQL queries to answer to CQ_3 and CQ_5 presented in Examples 3 and 4 are:

---

[6] http://www.w3.org/TR/owl-time/.
[7] World Wide Web Consortium. http://www.w3.org/.
[8] http://www6.inra.fr/cati-icat-atweb.
[9] http://www.wurvoc.org/vocabularies/om-1.8/.

Date	2015-06-28							
Project name	LIONES							
Microorganism (species, stem)	Saccharomyces cerevisiae - CBS 8066							
Code échantillon	2015-06-28-LIONES-001							

Step	SubStep	time harvest	Measure	Measure level	Scale	File number	Type of data
Fermentation	Preparation of the culture mixture	NA	Composition	Mixture		1	raw
Fermentation	Sterilisation of the culture mixture	NA	Conduite	Step		1	raw
Fermentation	Culture	24h	Conduite	Step		3	raw
Fermentation	Culture	24h	Cultivability	Mixture	Population	1	raw - computed
Fermentation	Cleaning	NA	Composition	Mixture	Mixture	1	raw
Concentration	NA	20 min	Conduite	Step		1	raw
Drying into stove	NA	75 min	Conduite	Step		2	raw
Drying into stove	NA	75 min	GSH intracellular	Mixture	Cellular	1	raw - computed
Drying into fluidized bed	60°60'	60 min	Conduite	Step		1	raw
Drying into fluidized bed	60°60'	60 min	GSH intracellular	Step	Cellular	1	raw - computed
Storage	NA	NA	Energy consumption	Step		1	computed

Fig. 7. An excerpt of the EXCEL file which describes a process

CQ_3:

SELECT ?prod (MAX(?value) AS ?valueMax) Where {

?attr rdf:type ms2o:Cultivability_losses .

?attr ms2o:hasForObservation ?obs .

?obs ms2o:hasForType ?type .

?type ms2o:hasForValue ?value .

?foi ms2o:hasForAttibute ?attr .

?foi ms2o:isComposedOf ?prod .

?prod rdf:type ms2o:Yeast .

?step ms2o:hasForMixture ?foi .

?step rdf:type ms2o:Drying

}

CQ_5:

Select ?obs Where {

?attr rdf:type ms2o:Drying_resistance .

?attr ms2o:hasForObservation ?obs .

?foi ms2o:hasForAttibute ?attr .

?foi ms2o:isComposedOf ?prod .

?prod rdf:type ms2o:Yarrowia_lipolytica .

?step ms2o:hasForMixture ?foi .

?step rdf:type ms2o:Fermentation

}

5   Conclusions and Further Work

In this paper, we presented the building of $[MS]^2O$, a Multi-Scale and Multi-Step Ontology for transformation processes, applied to the production of stabilized micro-organisms use case, performed at INRA (French National Institute for Agricultural Research). To build $[MS]^2O$, we have followed the NeOn methodology. We detailed the $[MS]^2O$ core component in which a transformation process is composed of steps, non necessarily in sequence, that allows inputs to be transformed in several different outputs. We stated that a transformation process relies on experimental observations that are performed either on a mixture (i.e. a set of combined products) at a certain scale during a certain step or on a step, the mixture being transformed during the steps of the process.

Future work is to define the module Material and Method and its connections with the other modules. We are currently investigating if the Semantic Sensor Network ontology[10] recommended by the W3C could be used. We are also evaluating how to enrich the product hierarchy with the FAO thesauri, AGROVOC[11] and the EFSA classification, FoodEx2[12].

In order to test the genericity of our $[MS]^2O$ core component modeling we are currently investigating how to use it in an other project concerning transformation processes on dairy gels.

$[MS]^2O$ ontology aims to play a key role as the representation layer of the querying and simulation systems of the LIONES project. This leads to the possibility of comparing different production scenarios and suggesting improvements while increasing efficiency (e.g. costs for the company). Moreover modeling the production processes of multi-scale objects may help domain experts to discover new semantic links between concepts and perform a cause effect analysis.

Acknowledgments. We are very grateful for the valuable inputs from all the domain experts partners involved in CellExtraDry French national project.

References

1. Béal, C., Marin, M., Fontaine, É., Fonseca, F., Obert, J.P.: Production et conservation des ferments lactiques et probiotiques. In: Georges, C., François-Marie, L. (eds.) Bactéries lactiques. De la génétique aux ferments, pp. 661–786. Lavoisier (2008)
2. Bizer, C.: Interlinking scientific data on a global scale. Data Sci. J. **12**, GRDI6–GRDI12 (2013)
3. Blomqvist, E., Hitzler, P., Janowicz, K., Krisnadhi, A., Narock, T., Solanki, M.: Considerations regarding ontology design patterns. Semant. Web **7**(1), 1–7 (2016)
4. Buche, P., Dibie-Barthélemy, J., Ibanescu, L., Soler, L.: Fuzzy web data tables integration guided by an ontological and terminological resource. IEEE Trans. Knowl. Data Eng. **25**(4), 805–819 (2013)
5. del Carmen Suárez-Figueroa, M., Gómez-Pérez, A., Fernández-López, M.: The NeOn methodology for ontology engineering. In: del Carmen Suárez-Figueroa, M., Gómez-Pérez, A., Motta, E., Gangemi, A. (eds.) Ontology Engineering in a Networked World, pp. 9–34. Springer, Heidelberg (2012)
6. Gautier, J., Passot, S., Pénicaud, C., Guillemin, H., Cenard, S., Lieben, P., Fonseca, F.: A low membrane lipid phase transition temperature is associated with a high cryotolerance of *Lactobacillus delbrueckii* subspecies bulgaricus CFL1. J. Dairy Sci. **96**(9), 5591–5602 (2013)
7. Grubic, T., Fan, I.S.: Supply chain ontology: review, analysis and synthesis. Comput. Ind. **61**(8), 776–786 (2010). Semantic Web Computing in Industry

---

[10] http://www.w3.org/2005/Incubator/ssn/XGR-ssn-20110628/.

[11] http://aims.fao.org/vest-registry/vocabularies/agrovoc-multilingual-agricultural-thesaurus.

[12] http://www.efsa.europa.eu/fr/datex/datexfoodclass.

8. Guarino, N., Oberle, D., Staab, S.: What is an ontology? In: Staab, S., Studer, R. (eds.) Handbook on Ontologies. International Handbooks on Information Systems, pp. 1–17. Springer, Heidelberg (2009)

9. Janowicz, K., Krisnadhi, A., Hu, Y., Suh, S., Weidema, B.P., Rivela, B., Tivander, J., Meyer, D.E., Berg-Cross, G., Hitzler, P., Ingwersen, W., Kuczenski, B., Vardeman, C., Ju, Y., Cheatham, M.: A minimal ontology pattern for life cycle assessment data. In: Blomqvist, E., Hitzler, P., Krisnadhi, A., Narock, T., Solanki, M. (eds.) Proceedings of the 6th Workshop on Ontology and Semantic Web Patterns (WOP 2015) Co-located with the 14th International Semantic Web Conference (ISWC 2015), Bethlehem, PA, USA, 11 October 2015. CEUR Workshop Proceedings, vol. 1461. CEUR-WS.org (2015)

10. Muljarto, A.-R., Salmon, J.-M., Neveu, P., Charnomordic, B., Buche, P.: Ontology-based model for food transformation processes - application to winemaking. In: Closs, S., Studer, R., Garoufallou, E., Sicilia, M.-A. (eds.) MTSR 2014. CCIS, vol. 478, pp. 329–343. Springer, Heidelberg (2014)

11. Pénicaud, C., Landaud, S., Jamme, F., Talbot, P., Bouix, M., Ghorbal, S., Fonseca, F.: Physiological and biochemical responses of *Yarrowia lipolytica* to dehydration induced by air-drying and freezing. PLoS ONE **9**, e111138 (2014)

12. Rospocher, M., Ghidini, C., Serafini, L.: An ontology for the business process-modelling notation. In: Garbacz, P., Kutz, O. (eds.) Proceedings of the Eighth International Conference on Formal Ontology in Information Systems, FOIS 2014, Rio de Janeiro, Brazil, 22–25 September 2014. Frontiers in Artificial Intelligence and Applications, vol. 267, pp. 133–146. IOS Press (2014)

13. Vardeman, C., Krisnadhi, A., Cheatham, M., Janowicz, K., Ferguson, H., Hitzler, P., Buccellato, A., Thirunarayan, K., Berg-Cross, G., Hahmann, T.: An ontology design pattern for material transformation. In: de Boer, V., Gangemi, A., Janowicz, K., Lawrynowicz, A. (eds.) Proceedings of the 5th Workshop on Ontology and Semantic Web Patterns (WOP 2014) Co-located with the 13th International Semantic Web Conference (ISWC 2014), Riva del Garda, Italy, 19 October 2014. CEUR Workshop Proceedings, vol. 1302, pp. 73–77. CEUR-WS.org (2014)

14. Velly, H., Bouix, M., Passot, S., Pénicaud, C., Beinsteiner, H., Ghorbal, S., Lieben, P., Fonseca, F.: Cyclopropanation of unsaturated fatty acids and membrane rigidification improve the freeze-drying resistance of *Lactococcus lactis* subsp. lactis TOMSC161. Appl. Microbiol. Biotechnol. **99**(2), 907–918 (2015)

Knowledge Engineering Method Based on Consensual Knowledge and Trust Computation: The MUSCKA System

Fabien Amarger[1,2], Jean-Pierre Chanet[1], Ollivier Haemmerlé[2],
Nathalie Hernandez[2], and Catherine Roussey[1(✉)]

[1] Irstea, UR TSCF Technologies et systèmes d'information pour les agrosystèmes,
9 Avenue Blaise Pascal, CS 20085, 63178 Aubiére, France
{fabien.amarger,jean-pierre.chanet,catherine.roussey}@irstea.fr
[2] Département de Mathématiques-Informatique, IRIT, UMR 5505, UT2J,
5 allées Antonio Machado, 31058 Toulouse Cedex, France
{ollivier.haemmerle,nathalie.hernandez}@univ-tlse2.fr

Abstract. We propose a method for building a knowledge base addressing specific issues such as covering end-users' needs. After designing an ontology module representing the knowledge needed, we enrich and populate it automatically with knowledge extracted from existing sources such as thesauri or classifications. The originality of our proposition is to propose ontological object candidates from existing sources according to their relatedness to the ontological module and to their trust score. This paper describes the trust measures we propose which are obtained by analysing the consensus found in existing sources. We consider that knowledge is more reliable if it has been extracted from several sources. Our measures has been evaluated on a real case study with experts from the agriculture domain.

Keywords: Ontology development · Trust · Non-ontological sources · Ontology Design Pattern · Ontology merging

1 Introduction

In many fields, domain specific information is distributed on the Web as structured data (such as databases or thesauri) gathered for a specific usage. End-users are often lost when facing this amount of data as they have to look for available sources, analyse their quality, retrieve specific information from each of them and compare them. Alongside, the Linked Open Data (LOD) initiative aims at linking and facilitating querying on available data. Approaches such as [18,20] have been proposed to formalise existing sources, define vocabularies to describe them and publish them on the LOD. However, approaches are still needed in order to help end-users collect and access knowledge to achieve a specific task in specialised domains. We proposed a method for building a Knowledge Base (KB) covering end-users' needs by extracting knowledge from non-ontological

O. Haemmerlé et al. (Eds.): ICCS 2016, LNAI 9717, pp. 177–190, 2016.
DOI: 10.1007/978-3-319-40985-6_14

sources such as thesauri or classifications. The main idea is to design an ontology module representing the knowledge needed by end-users and to populate it and enrich it automatically with data extracted from existing sources. The originality of our proposition is to identify ontological object candidates from existing sources according to both their relatedness to the ontological module and to a trust score. This trust is computed by analysing the consensus found in existing sources. Thus we consider that knowledge is more reliable if it has been extracted from several sources. A first experiment carried out with agronomic experts has validated the relevancy of our approach [4]. This experiment focused in extracting instances and properties between instances. As non-ontological sources often contain rich lexical data we propose to improve our method and add labels to the previous ontological objects. A deep analysis of existing sources has shown that we can improve our method and generate new classes, which specialised the ontological module classes and extract $rdf : type$ property between instances and classes. This paper is an extension of our previous work and presents measures to compute trust for each new type of ontological objects that are extracted by our method[1]. These measures are evaluated on a real case study in agriculture and we believe they can be applied in any knowledge base merging process.

The paper is organised as follows. First we give an overview on the MUSKCA system in order to explain how ontological object candidates are extracted from non-ontological sources. In Sect. 3, we present our proposal for computing consensual trust measures. Finally we describe and analyse the experiments.

2   Overview on the MUSKCA Approach

Our method is composed of three processes detailed in [4].

1. **Source Analysing:** During this process, the domain expert and the ontologist work together to select the most appropriate sources to build their KB. They inspect each source to evaluate its coverage and to have a broad idea if the source can be transformed to a KB or not.
2. **Source Transformation:** This process transforms each source into a KB in OWL format. It is based on Neon methods and consists in using transformation patterns for enriching and populating an ontological module defining a users' information need. The module is composed of $owl : Classes$ and defines the set of $owl : Properties$ that may exist between them. It is designed using Ontology Design Patterns and vocabularies already published on the LOD. An example of one of our modules is AgronomicTaxon [17]. At the end of this stage, the automatically generated KBs are composed of specialisation and instantiation of the classes and properties defined in the ontological module. Figure 1 shows a sample of the automatically generated KB for the AgronomicTaxon module using Agrovoc.
3. **KB Merging:** This process builds the final KB based on all KBs extracted from sources. As far as we know, this process is not proposed in any ontology

---

[1] More information about the extraction method is available in french language [2,3].

engineering method. Usually an ontology engineering method uses several sources separately in order to enrich the KB in an incremental way. Here, the merging process uses several KBs at the same time in order to extract consensual knowledge.

This paper focuses on the measures used during the last step in order to merge ontological objects extracted from the different KBs. Ontological objects can be of several kinds:

- subclasses of classes of the ontological module (for example *Subkingdom* in Fig. 1)
- instances of classes or subclasses of the ontological module (*Plantae* in Fig. 1)
- labels for these classes and instances (Not represented on Fig. 1 for clarity)
- rdf:type property linking instances to classes (the property between *Plantae* and *Neon : Kingdom* in Fig. 1)
- Any property defined in the ontological module that links instances (the relation *hasHigherRank* between *Embryophyta* and *Plantae* in Fig. 1

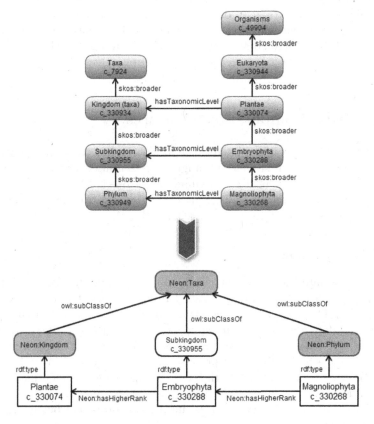

Fig. 1. Examples of ontological objects extracted from Agrovoc

Extracting ontological objects from various sources requires considering the trust that can be given to these objects. Several definitions of the notion of trust in the fields of computer science and semantic web are presented in [5]. The one which corresponds the most to our purpose is:

"Trust of a party A to a party B for a service X is the measurable belief of A in that B behaves dependable for a specified period within a specified context (in relation to service X)."

We consider A as the user who wants to create a knowledge base, B as a source and X the extraction process, for a specific period and context.

In Information Retrieval and Information Extraction, a common assumption is that the frequency of a word or a phrase increases the trust that can be given to this piece of information [8]. On the web, the reputation of pages is also taken into consideration by computing trust using the hypertext links they contains, as done in the well known PageRank algorithm. Authors of [7] claim that in Semantic Web there is more to trust than reputation, putting notably forwards that the context in which a statement occurs has to be considered. This position has motivated the fact that we consider building a KB corresponding to a specific users' need represented by an ontological module. [10] highlights the fact that measures are needed in order to assign trust to specific pieces of a source, considering that when evaluating trust the information contained in the source can not be considered as a whole. As many propositions have been made in order to measure the trust to give to the sources themselves [6], this paper focuses on trust measures to order KB candidate.

3   Our Trust Computation Measures

To compute the trust that can be given to an ontological object found in several sources, we propose to first reuse an alignment tool in order to identify equivalent elements that we consider as redundant (Subsect. 3.1). Then we generate potential candidates (Subsect. 3.2). Finally we propose two ways of computing the trust for these candidates (Subsects. 3.3 and 3.4).

3.1   Using Mappings

In order to evaluate the trust of a specific object in a given KB, we propose to consider mapping that can be established from this object to other objects of different KBs. This idea follows the same principle as exploiting hyperlinks to compute trust for web pages. As mappings between objects of different KB have to be identified, we reuse alignment techniques. Alignment is a large research area [9] and many methods have been proposed and implemented in tools. For the moment, mature tools only deal with homogeneous mappings identified between 2 knowledge bases [11]. We thus consider, in this paper, mappings between ontological objects of the same type (class, individual, property) belonging to two knowledge bases.

Let's consider two knowledge bases KB_1 and KB_2. Aligning KB_1 and KB_2 consists in computing all the mappings between objects of KB_1 and objects of KB_2 of the same type.

Let's define a mapping m as a triplet $<e_i, e_j, s_{ij}>$ such as:

- $e_i \in KB_i$: is an ontological object belonging to KB_i,
- $e_j \in KB_j$: is another ontological object belonging to KB_j ($KB_j \neq KB_i$),
- $type(e_i) = type(e_j)$: e_i and e_j are ontological objects of the same type (class, individual, object property, etc.),
- $s_{ij} = degree(e_i, e_j)$: where $degree$ is a function from $KB_i \times KB_j$ to $[0, 1]$ giving the similarity degree computed by an alignment tool for e_i and e_j.

Our goal is to explicitly identify the redundancy between sources, that is to say the ontological objects that correspond to each other in the different sources (at least 2, but possibly more). To do so, we consider that all the mappings between each possible pair of knowledge bases are generated.

3.2 Candidate Generation

According to the generated mapping between each possible pair of KB, we group similar objects in what we call a candidate. We define $dim(c)$ as the number of KBs involved in a candidate c. We identify five kinds of candidates:

Class Candidate (cc). We define a class candidate cc as a set of mappings associating pairs of classes belonging to different knowledge bases. The set of mappings identifies similar classes in the different KBs. For n knowledge bases, a class candidate will be composed, at the most, of $n * (n - 1)/2$ mappings.

Individual Candidate (ic). Individuals are instances of classes. In the same way as for class candidates, we identify an individual candidate ic as a set of mappings associating pairs of instances belonging to different knowledge bases. Let's consider the example in Fig. 2 with three knowledge bases KB_1,

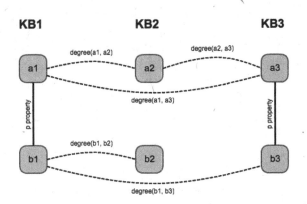

Fig. 2. Example of individual and relation candidates

KB_2 and KB_3. KB_1 and KB_3 contain two individuals a_i, b_i linked by the same property p. The dash lines represent mappings between individuals generated by the alignment tool. There are two individual candidates ic_1 and ic_2. In this example, $dim(ic_1) = 3$, $dim(ic_2) = 3$ and $dim(rc_1) = 2$.

$$ic_1 = \begin{cases} [<a_1, a_2, s_{12}>, <a_2, a_3, s_{23}>, <a_1, a_3, s_{13}>] \\ a_1 \in KB_1, a_2 \in KB_2, a_3 \in KB_3 \\ s_{12} = degree(a_1, a_2) \\ s_{23} = degree(a_2, a_3) \\ s_{13} = degree(a_1, a_3) \end{cases}$$

$$ic_2 = \begin{cases} [<b_1, b_2, s_{12}>, <b_1, b_3, s_{23}>] \\ b_1 \in KB_1, b_2 \in KB_2, b_3 \in KB_3 \\ s_{12} = degree(b_1, b_2), s_{13} = degree(b_1, b_3) \end{cases} \qquad (1)$$

Relation Candidate (rc). We define a relation candidate rc as a pair of individual candidates, such as there exists the same property that links components of individual candidates of the different KB. For example in Fig. 2 the p property links $a1$ and $b1$ and also links $a3$ and $b3$. $rc1$ is a candidate relation.

$$rc_1 = \begin{cases} [ic_1, ic_2] \\ [p(a_1, b_1), p(a_3, b_3)] \\ p(a_1, b_1) \in KB_1, p(a_3, b_3) \in KB_3 \end{cases} \qquad (2)$$

Type Candidate (tc). $rdf : type$ properties link an individual candidate to its type. Its type can either be a class candidate or a class that already exists in the module. Thus there exist two kinds of type candidates depending on the

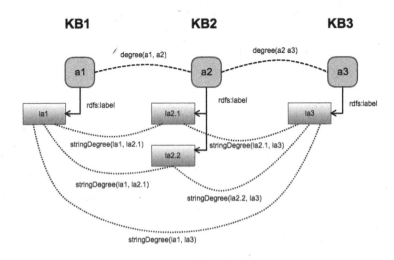

Fig. 3. Example of label candidates

target of the $rdf : type$ property. We define a type candidate tc as an instance candidate that is linked to its type, such as there exists some $rdf : type$ properties that link a component of the individual candidate to a component of the class candidate or a class that is defined in the module.

Label Candidate (lc). A label candidate lc is associated to a class candidate cc or to an individual candidate ic, called the *root* of lc. The root should have labels for at least two of its components. A new string mapping function is processed on the labels belonging to distinct components. We define a label candidate thanks to these new string mappings. A label candidate lc is a set of string mappings that links labels of distinct components of the root. For simplicity dim of a label candidate lc will be equal to dim of its root. Let's consider Fig. 3 which represents 3 individuals and associated labels extracted from three knowledge bases $KB1$, $KB2$ and $KB3$. The dashed lines annotated by $degree(a_i, a_j)$ correspond to mapping generated by the alignment tool and the corresponding degree. The dashed lines annotated by $stringDegree(lab_i, lab_j)$ correspond to the mappings generated by the string mapping function. In this figure there are two examples of label candidates having the same root (the shared individual candidate). The root contains three individuals $a1$, $a2$, $a3$. $a1$ has one label $la1$. $a2$ has two labels $la2.1$ and $la2.2$. $a3$ has also one label $la3$.

$$root = \begin{cases} [<a_1, a_2, s_{r1}>, <a_2, a_3, s_{r2}>, <a_1, a_3, s_{r3}>] \\ a_1 \in KB_1, a_2 \in KB_2, a_3 \in KB_3 \\ s_{r1} = degree(a_1, a_2) \\ s_{r2} = degree(a_2, a_3) \\ s_{r3} = degree(a_1, a_3) \end{cases}$$

$$lc_1 = \begin{cases} [<la_1, la_{2.1}, s_{11}>, <la_{2.1}, la_3, s_{12}>] \\ la_1 \in KB_1, la_{2.1} \in KB_2, la_3 \in KB_3 \\ s_{11} = stringDegree(la_1, la_{2.1}), \\ s_{12} = stringDegree(la_{2.1}, la_3) \end{cases} \tag{3}$$

$$lc_2 = \begin{cases} [<la_1, la_{2.2}, s_{21}>, <la_{2.2}, la_3, s_{22}>] \\ la_1 \in KB_1, la_{2.2} \in KB_2, la_3 \in KB_3 \\ s_{21} = stringDegree(la_1, la_{2.2}), \\ s_{22} = stringDegree(la_{2.2}, la_3) \end{cases}$$

Each candidate has a trust score to define how much we can trust this candidate. There are several ways to compute this score. We define several trust functions that we will test in the experiments.

3.3   Simple Trust Function

A simple way to extract consensual ontological objects is to determine in how many KBs the candidate appears. We compute a ratio between the $dim(c)$ and the $nb_{sources}$ (the total number of sources used). We defined a function called $trust_{simple}$ to implement simple consensus:

$$trust_{simple}(c) = \frac{dim(c)}{nb_{Sources}} \tag{4}$$

3.4 Degree Trust Function

With the previous function, we consider that all the mappings proposed by the alignment tool are correct and can be trusted in the same way. With the following measures, we take into consideration both the number of sources in which the ontological object has been found and the number of mappings that have been established for each object. The intuition is that the more mappings can be established from objects of distinct sources the more it can be trusted. The measures also take into consideration the mapping degree computed by the alignment tool. Thus, the more potential mappings (even if not sure) can be established the better is the candidate.

For the degree consensus implementation there is a different formula for each kind of candidate.

Individual Candidate Trust Degree Function. This function is defined by the formula:

$$trust_{degree}(ic) = \frac{\sum_{i=1}^{dim(ic)} \sum_{j=i+1}^{dim(ic)} degree(a_i, a_j)}{\frac{nb_{Sources}(nb_{Sources}-1)}{2}} \tag{5}$$

$$such\ as(a_i, a_j) \in ic$$

This function sums all the mapping degrees involved in the candidate. We normalised the result with the maximum number of individual mappings possible in an individual candidate (since we have 3 KBs thus we can have at most 3 mappings in an individual candidate). Here, $nb_{sources}$ is the total number of KBs involved in the merging process.

Class Candidate Trust Degree Function. This function is defined by the formula:

$$trust_{degree}(cc) = \frac{\sum_{i=1}^{dim(cc)} \sum_{j=i+1}^{dim(cc)} degree(c_i, c_j)}{\frac{nb_{Sources}(nb_{Sources}-1)}{2}} \tag{6}$$

$$such\ as(c_i, c_j) \in cc$$

This function is the same as the individual candidate trust function except that it considers the classes.

Relation Candidate Trust Degree Function. This trust function is defined by the formula:

$$trust_{degree}(rc) = \frac{dim(rc) + \frac{trust(ic1)+trust(ic2)}{2}}{nb_{Sources} + 1} \tag{7}$$

$$\text{such as } ic1 \in rc, ic2 \in rc$$

This formula takes into account the $dim(rc)$ and the average of the trust scores of individual candidates, components of the relation candidate. We do so to simulate a mapping degree between object properties as alignment tools do not match object properties. We normalise this result with the $nb_{Sources}$, which is the maximum value that $dim(rc)$ could be, plus 1, which is the maximum value that the average of the two ic trust scores could be.

Type Candidate Trust Degree Function. There exist two trust functions for type candidate:

$$trust_{degree}(tc_1) = \frac{dim(tc_1) + \frac{trust(ic1)+trust(cc2)}{2}}{nb_{Sources} + 1} \tag{8}$$

$$\text{such as } ic1 \in tc_1, cc2 \in tc_1$$

Formula 8 is dedicated to type candidate tc_1 that is composed of an individual candidate $ic1$ and a class candidate $cc2$. This formula takes into account the $dim(tc_1)$ and the average of trust scores of individual candidate and class candidate, components of the type candidate. We do so to simulate a mapping degree between properties. We normalise this result as in formula 7

$$trust_{degree}(tc_2) = \frac{dim(tc_2) + trust(ic)}{nb_{Sources} + 1} \tag{9}$$

$$\text{such as } ic \in tc_2$$

Formula 9 is dedicated to type candidate tc_2 that is composed of an individual candidate ic and a class that already exists in the module.

Label Candidate Trust Degree Function. This function is defined by the formula:

$$trust_{degree}(lc) = \frac{trust_{degree}(root) + sum\_string\_degree(lc)}{2}$$

$$sum\_string\_degre(lc)$$

$$= \frac{\sum_{i=1}^{dim(root)} \sum_{j=i+1}^{dim(root)} stringDegree(label(a_i), label(a_j))}{\frac{nb_{Sources}(nb_{Sources}-1)}{2}} \tag{10}$$

$$\text{such as } a_i, a_j \in root$$

$$label(a_i), label(a_j) \in lc$$

This formula uses a label function called *label*() that returns the label of an individual or a class, component of the root of *lc*. The *sum_string_degree*() function sums the degrees of string mappings involved in the label candidate. We normalise the result with the maximum number of possible string mappings for candidate labels. The trust score of a label candidate sums the trust score of its root and the *sum_string_degree* value. We normalise the result, because its component takes its value between zero and one.

4   Experiments

Our approach has been implemented in a prototype called MUSKCA developed in Java. It is available on github at https://github.com/Murloc6/Muskca. After analysing the results of the OAEI challenge[2] and especially the ones of the instance matching task[3] [11] dealing with tools mapping all kinds of ontological objects, we chose to use LogMap [13] of which the source code is available online[4].

In our experiments, we used MUSKCA on a real case study for which a knowledge base about plant classification is needed. This knowledge base will be used for analysing and annotating alert bulletins that inform farmers of pest attacks on crops. The specific information need is represented in the ontological module AgronomicTaxon described in [17]. Three well known sources: Agrovoc, Taxref, NCBI were considered in order to generate the knowledge base enriching and populating the AgronomicTaxon module. As our aim is to evaluate to what extent the trust measures help identifying relevant ontological objects from existing sources, we decided to compare the candidates identified thanks to the consensual trust score computation with the ontological objects manually selected from each source by the experts.

4.1   Gold Standard

We asked three domain experts to analyse the three knowledge bases extracted automatically from the three sources. The experts had to determine for each source which ontological objects were relevant and if they are in the scope of the ontological module. An interface was implemented to collect the experts' opinion. Note that a real effort was made in order to present ontological objects in a way understandable for experts. Here are some questions asked:

- **Does *Magnoliophyta* belong to the domain?** We want to know if the instance Magnoliophyta is relevant and in the scope of the KB.
- **Does *angiosperm* designate *Magnoliophyta*?**
  We want to know if the labels associated to Magnoliophyta are correct as they are sometimes inexact (most of the time not synonyms) or not the right translation (if the source contains multilingual labels).

---

[2] Ontology Alignment Evaluation Initiative - http://oaei.ontologymatching.org/2013/.

[3] http://www.instancematching.org/oaei/imei2013/results.html.

[4] https://code.google.com/p/logmap-matcher/.

Table 1. Results of the first experiment: $trust_{simple}$ and threshold permissive

Candidate type	Precision	Recall	F-Measure
Individual	0.92	0.66	0.77
Relation	0.65	0.51	0.57
Type	0.70	0.43	0.54
Label	0.32	0.35	0.34
Class	1	0.38	0.55

We thus obtained a list of ontological objects validated for each source. We then compared them with the candidates generated by the prototype. To do that we computed the precision, recall and f-measure for each kind of candidate.

4.2 Results

We ran two experiments:

1. The first experiment uses the $trust_{simple}$ function to compute the trust score of the candidates and a relatively permissive threshold for filtering candidates (fixed at 0.5 which means that ontological objects are found in at least half of the sources).
2. The second experiment uses the $trust_{degree}$ function to compute the trust score of the candidates and a threshold relatively permissive for filtering candidates (fixed at 0.6 in order to compare the results of these measures on the same amount of candidates as in the first experiment)

Tables 1 and 2 present the results of experiments on each kind of candidates.

As we can see in Table 1, filtering candidates if they appear in at least half of the sources through the $trust_{simple}$ measure helps identifying relevant candidates. The relatively significant precision we obtain validates our intuition that redundant objects identified through mappings are a way of finding and ranking relevant candidates. Note that the results are less meaningful on labels. This can be explained by the fact that only one source (Agrovoc) contains lots of labels. The two others have one label by ontological object. Recall results show that consensual trust is not the only approach that should be used in order to extract all the candidate from sources but at least those extracted are relevant.

Table 2 presents the results of the second experiment carried out with the $trust_{degree}$ measures exploiting the number of mappings and their degree. As we can see the precision increases. This is due to the fact that candidates that group 3 similar objects linked by 3 mappings are better ranked than candidates that group 3 similar objects linked by 2 mappings. We believe that the improvement would be more significant in an experiment involving more than 3 sources. Indeed in our experiment the number of mappings for an object vary from 2 to 3. With more than 3 sources, we hope that the number of mappings will vary to a larger extent.

During these experiments we compared two ways of computing a trust score on ontological objects based on their consensual degree. The comparison of the Tables 1 and 2 shows that the use of $trust_{degree}$ ranks relevant candidates better. This happens because the consensual aspect has more impact in this formula than in $trust_{simple}$. This first experiment shows that the use of the consensus in the trust score computation increases the quality of the results.

Table 2. Results of the second experiment: $trust_{degree}$ and threshold permissive

Candidate type	Precision	Recall	F-Measure
Individual	0.97	0.63	0.76
Relation	0.68	0.51	0.58
Type	0.77	0.39	0.52
Label	0.39	0.24	0.30
Class	1	0.13	0.23

5   Related Works

Our work on trust computation can be compared to the work of [15] that computes a trust score for mappings generated by several alignment tools. They propose a fuzzy voting model supported by alignment agents. The fuzzy model helps to compare the numerical results. The trust assigned to each agent depend on the alignment situation. For example, an agent can have the best result when aligning labels, but it can have the worst result for aligning instances. Our work can be seen has a voting between several sources in order to decide if the ontological object will appear in the final KB.

Several approaches have been proposed in order to correct mapping by identifying contradictory elements generated by alignment tools [1,12,14,16,19]. All these approaches deals the alignment of only two ontologies. They detect conflicts between a set of mappings. Their goal is to remove a minimal subset of mappings in order to keep coherent the alignment of the two ontologies. As formalised in [14] when two ontologies are aligned, the resulting ontology contains doubloons classes. In this paper, our goal is to focus on consensual knowledge identified between the sources and create only one ontological object by candidate.

6   Conclusion and Future Works

Our work consists in building a knowledge base with several kinds of ontological objects (individual, relation instance, type relation, label and classes) extracted from non-ontological sources. In this paper, we proposed two ways to compute the consensual trust score for filtering the potential candidates extracted from the sources. The first formula, $trust_{simple}$ is the ratio between the number of sources in which the candidate appears and the total number of sources considered. The second formula, $trust_{degree}$ takes into account, for each similar object in the different sources, the degree of all the mappings given by an alignment system. This formula gives more weight to consensus and to the quality of the agreement between the sources. An experiment involving experts from the agriculture domain has shown that the use of consensus in the trust score computation increases the quality of the results. We are currently defining a new evaluation protocol in order to analyse more deeply our approach when more than 3 sources are considered. The low precision obtained for the label candidates can be explained by the lack of consensus for this kind of ontological object in the sources we considered. The solution to this problem could be to emphasise the potential strength of each source in our process. In the context of our experiment, before analysing in depth the sources, the experts believed that labels extracted from Agrovoc were going to be more relevant than labels from NCBI. Approaches such as [6] could be used to evaluate beforehand the strengths of the considered sources. In this paper we consider all candidates independently from one another. We are aware that some candidates can be contradictory. Next we plan to propose an approach to select a set of candidates without contradictions. We would like to reuse works such as [15] to deal with contradictory representations.

Acknowledgments. We want to thank specially the three experts who helped us: Franck Jabot and Vincent Soulignac from Irstea Clermont-Ferrand and Jacques Le Gouis from INRA Clermont-Ferrand.

References

1. Abbas, M.A., Berio, G.: Creating ontologies using ontology mappings: compatible and incompatible ontology mappings. In: Web Intelligence (WI) and Intelligent Agent Technologies (IAT), pp. 143–146 (2013)
2. Amarger, F.: Vers un systeme intelligent de capitalisation de connaissances pour l'agriculture durable: construction d'ontologies agricoles par transformation de sources existantes. Ph.D. thesis, Université de Toulouse 2 le Mirail (2015)
3. Amarger, F., Chanet, J.-P., Haemmerlé, O., Hernandez, N., Roussey, C.: Construction d'une ontologie par transformation de systèmes d'organisation des connaissances et évaluation de la confiance. Ingénierie des Systèmes d'Information **20**(3), 37–61 (2015)
4. Amarger, F., Chanet, J.-P., Haemmerlé, O., Hernandez, N., Roussey, C.: SKOS sources transformations for ontology engineering: agronomical taxonomy use case. In: Closs, S., Studer, R., Garoufallou, E., Sicilia, M.-A. (eds.) MTSR 2014. CCIS, vol. 478, pp. 314–328. Springer, Heidelberg (2014)
5. Artz, D., Gil, Y.: A survey of trust in computer science and the semantic web. Web Semant. Sci. Serv. Agents World Wide Web **5**, 58–71 (2007)

6. Balakrishnan, R., Kambhampati, S.: Sourcerank: relevance and trust assessment for deep web sources based on inter-source agreement. In: Proceedings of the 20th International Conference on World Wide Web, WWW 2011, pp. 227–236. ACM, New York (2011)

7. Bizer, C., Oldakowski, R.: Using context- and content-based trust policies on the semantic web. In: Proceedings of the 13th international conference on World Wide Web - Alternate Track Papers & Posters, WWW 2004, 17–20 May 2004, pp. 228–229. ACM, New York (2004)

8. Clarke, C.L.A, Cormack, G.V., Lynam, T.R.: Exploiting redundancy in question answering. In: SIGIR 2001: Proceedings of the 24th Annual International Conference on Research and Development in Information Retrieval, pp. 358–365. ACM, September 2001

9. Euzenat, J., Shvaiko, P.: Ontology Matching, vol. 333. Springer, Heidelberg (2007)

10. Gil, Y., Artz, D.: Towards content trust of web resources. In: Proceedings of the 15th International Conference on World Wide Web, WWW 2006, Edinburgh, Scotland, UK, 23–26 May 2006, pp. 565–574 (2006)

11. Grau, B.C., Dragisic, Z., Eckert, K., Euzenat, J., Ferrara, A., Granada, R., Ivanova, V., Jiménez-Ruiz, E., Kempf, A.O., Lambrix, P., et al.: Results of the ontology alignment evaluation initiative 2013. In: ISWC Workshop on Ontology Matching (OM), pp. 61–100 (2013)

12. Jean-Mary, Y.R., Kabuka, M.R.: ASMOV: results for OAEI 2008. In: Proceedings of the 3rd International Workshop on Ontology Matching (OM 2008), Karlsruhe, Germany, vol. 431. CEUR-WS.org, October 2008

13. Jiménez-Ruiz, E., Grau, B.C., Zhou, Y., Horrocks, I.: Large-scale interactive ontology matching: algorithms and implementation. In: European Conference on Artificial Intelligence, pp. 444–449 (2012)

14. Meilicke, C., Stuckenschmidt, H.: An efficient method for computing alignment diagnoses. In: Polleres, A., Swift, T. (eds.) RR 2009. LNCS, vol. 5837, pp. 182–196. Springer, Heidelberg (2009)

15. Nagy, M., Vargas-Vera, M.: Dealing with contradictory evidence using fuzzy trust in semantic web data. In: Bobillo, F., Costa, P.C.G., d'Amato, C., Fanizzi, N., Laskey, K.B., Laskey, K.J., Lukasiewicz, T., Nickles, M., Pool, M. (eds.) URSW 2008-2010/UniDL 2010. LNCS, vol. 7123, pp. 139–157. Springer, Heidelberg (2013)

16. Raunich, S., Rahm, E.: Target-driven merging of taxonomies with Atom. Inf. Syst. **42**, 1–14 (2014)

17. Roussey, C., Chanet, J.-P., Cellier, V., Amarger, F.: Agronomic taxon. In: WOD, p. 5 (2013)

18. Soergel, D., Lauser, B., Liang, A., Fisseha, F., Keizer, J., Katz, S.: Reengineering thesauri for new applications: the AGROVOC example. J. Digit. Inf. **4**, 1–23 (2004)

19. Trojahn, C., Euzenat, J., Tamma, V., Payne, T.R.: Argumentation for reconciling agent ontologies. In: Elçi, A., Koné, M.T., Orgun, M.A. (eds.) Semantic Agent Systems. SCI, vol. 344, pp. 89–111. Springer, Heidelberg (2011)

20. Villazón-Terrazas, B., Suárez-Figueroa, M.C., Gómez-Pérez, A.: A pattern-based method for re-engineering non-ontological resources into ontologies. Int. J. Semant. Web Inf. Syst. **6**, 27–63 (2010)

Web-Mining Defeasible Knowledge from Concessional Statements

Alina Petrova[1]([✉]) and Sebastian Rudolph[2]

[1] University of Oxford, Oxford, UK
alina.petrova@cs.ox.ac.uk
[2] Technische Universität Dresden, Dresden, Germany

Abstract. Mining common-sense knowledge is a vital problem of artificial intelligence that forms the basis of various tasks, from information retrieval to robotics. There have been numerous initiatives to mine common-sense facts from unstructured data, more specifically, from Web texts. However, common-sense knowledge is typically not explicitly stated in the text, as it is considered to be obvious, self-evident, and thus shared between writer and reader. We argue that certain types of *defeasible* common-sense knowledge (i.e., knowledge that holds in most but not all cases), in particular, beliefs and stereotypes, tend to appear in text in a particular manner: they are not explicitly manifested, unless the speakers encounter a situation that runs in contrast to their defeasible common-sense assumptions. For example, if a speaker believes that Spain is a very warm country, she may express a surprise when it snows in Bilbao. We further argue that such conceptual contradictions correspond to the linguistic relation of concession (e.g., *although Bilbao is in Spain, it is snowing there today*) and we present a methodology for extracting defeasible common-sense beliefs (*it is not common to snow in Spain*) from Web data using concessive linguistic markers. We illustrate the methodology by mining beliefs about persons and we show that we are able to extract new information compared to existing common-sense knowledge bases.

1 Introduction

Common-sense knowledge is a set of basic propositions of a very broad semantics that describe different classes and instances, their most common properties (e.g., shape, color, material, frequency, age) and how they relate to each other. For example, the following statements belong to the realm of common-sense knowledge: *snow is white, London is in England, July is the seventh month of a year, children often believe in Santa Claus.* From a human-oriented point of view, "a common-sense fact is a true statement about the world that is known to most humans" [13], while from the point of view of formal systems, a common-sense fact is a formalized statement about the world that is shared between all agents and is true across all applications.

Mining common-sense knowledge from a variety of resources is a vital problem of artificial intelligence, it is required for common-sense reasoning which

© Springer International Publishing Switzerland 2016
O. Haemmerlé et al. (Eds.): ICCS 2016, LNAI 9717, pp. 191–203, 2016.
DOI: 10.1007/978-3-319-40985-6_15

forms the basis for a plethora of tasks, from more applied ones, such as question answering and item recommendation, to more general ones, such as intelligent decision making, robotics and natural language understanding [1].

For example, let us consider recommender systems. Having information about a given user, e.g., his age, profession or personal traits, the system could use common-sense knowledge about typical preferences of users with the same characteristics and generate additional item suggestions to the user, complementing the ones based on data mining and correlation. Suppose a user states that he is energetic, yet he has not viewed or purchased any sport-related articles. A knowledge base containing a statement *energetic people tend to enjoy sports* could expand the range of items recommended to the user, thus potentially benefiting both the user and the system and overcoming the cold-start problem [6].

Yet, the acquisition and deployment of common-sense knowledge in practical scenario is still underdeveloped: "the lack of common-sense knowledge and reasoning was encountered in many if not all application areas of Artificial Intelligence" [2].

There have been numerous initiatives to collect common-sense facts and to represent them formally, e.g., in the form of RDF triples [7,12,13]. The facts are usually mined from unstructured data, more specifically, from Web texts. This requires, however, that common-sense knowledge has to be explicitly stated in the texts, which tends to be not the case. On the contrary: as common-sense knowledge is considered to be obvious, self-evident and shared between writer and reader anyway, it is often not stated since there is no need to convey it. This comes as a consequence of Grice's conversational maxims of quantity and manner: when communicating, one tries to be as brief and concise as possible, and does not contribute more information than is actually required [3]. Humans are barely aware of the plethora of shared common-sense knowledge their communication is implicitly based on. As Davis put it: "Since common sense consists (by definition) of knowledge and reasoning methods that are utterly obvious to us, we often overlook its astonishing scope and power" [1].

We argue that, therefore, certain types of *defeasible* common-sense knowledge (i.e., knowledge that holds in most but not all cases), in particular, beliefs and stereotypes, tend to appear in text only in an indirect manner: they are not explicitly manifested, unless the speaker encounters a situation that runs in contrast to her defeasible common-sense assumptions. For example, if a speaker believes that Spain is a very warm country, she may express a surprise when it snows in Bilbao. We further argue that this type of conceptual contradictions closely corresponds to the linguistic relation of concession (e.g., *although Bilbao is in Spain, it is snowing there today*). In this paper, we present a methodology and proof-of-concept evaluation for extracting defeasible common-sense beliefs (such as *it is not common to snow in Spain*) from Web data using concessive linguistic markers. We illustrate the methodology by mining beliefs about persons and we show that we are indeed able to extract new information complementing the one in existing common-sense knowledge bases.

2   Preliminaries

Before presenting our methodology of extracting defeasible common-sense beliefs, we provide the cognitive and linguistic justification for our general approach by looking into the nature of defeasible knowledge and the semantics of concession.

2.1   Defeasible Knowledge

An argument is called *defeasible* if it is rationally compelling but not deductively valid [5]. In deductive reasoning, a statement *if p, then q* imposes a constraint that the conclusion q must be true if the premise p is true. In defeasible reasoning, the conclusion is believed to be true given p, but it can potentially be defeated by some additional argument, while p remains to be true, which reflects the non-monotonic nature of defeasible knowledge [9]. We can formulate a defeasible conditional statement in the following way: q is commonly believed to follow from p, although in certain situations this consequence relation may not hold, or in short: *if p, then* normally q. Hence, defeasible reasoning may be used to model common beliefs or stereotypes: we believe that q follows from p, but this is not formally proven and may have exceptions [8]. Defeasible knowledge may or may not be supported by empirical evidence.

2.2   Concession

In natural language, there exist numerous discourse relations. A well-studied relation that is commonly encountered in texts is the *opposition relation* [4]. It is a relation with broad semantics that links two contrasting, mutually exclusive items. A particular subtype of opposition relation is *concessive relation*, or *concession*. This relation links two potentially or apparently contrasting items due to an implicit assumption, known also as *default implication*: through this assumed implication, one argument creates an expectation which is then denied in the second argument [10]. For example, "although it is summer, the weather is not warm" is a concessive sentence which relies on the common assumption that it is warm in summer (with respect to the location of the speaker). The fact that it is summer, stated in the first argument, triggers the expectation of warm weather which is then refuted by the second argument. The assumption/expectation may follow logically from the first argument (*Jack got cold* (hence it is assumed he felt it), *but he did not realize it*), statistically correlate with it (*Nick did not buy any beer, although he was going to watch football with his friends* (and it is very common to have a beer in this situation)) or be based on a property that is commonly related to it (*although she is blonde* (and the stereotype is blonde women are not particularly smart), *she has a degree in biochemistry*).

Winter and Rimon [14] formalized the semantics of concession in the following way, $\Diamond$ being the possibility operator from classical modal logic: $(p \wedge q) \wedge \Diamond (p \rightarrow \neg q)$, where p is the first argument, q is the second argument and $\neg q$ is the expectation that was triggered by p. As we can see, the semantics of concession is very much aligned with that of a defeasible implication, and in fact, the default assumption that underlies concessive relation ($p \rightarrow \neg q$) is sometimes called *the defeasible rule.* Therefore, in order to find linguistic representation of common-sense knowledge that is manifested in text in a defeasible way, we will focus on concessive sentences:

▷ linguistic concession:
although p, q

▷ defeasible rule that is implicitly present in concession:
if p, then usually not q

2.3   Concessive Markers

In order to identify and extract concessive statements, it is helpful to determine what are the means of representation of concession in text. There are numerous linguistic studies of concession that analyse types of concessive relations and how they can be expressed in English. For instance, Taboada and Gómez-Gonzáles [11] present an extensive analysis of concessive relation and its *discourse markers* (elements of text that explicitly signal a particular relation). They study both spoken and written texts, in particular, book and movie reviews collected from the Web, which is in line with our setting, as we aim at analysing Web texts.

Concession can be expressed in natural language in multiple ways [11], namely, using subordinate conjunctions, coordinate conjunctions, adverbial items, phrasal expressions, and parenthetical elements. The discourse markers that accompany these means of representation are:

- conjunctions: *although, but, despite the fact that, even though* etc.;
- sentence adverbials: *nevertheless, regardless, yet* etc.;
- gerund constructions: *supposing, granting* etc.;
- prepositions: *in spite of, regardless of* etc.

Despite the heterogeneity of concessive means of representation, the vast majority of them are very rarely encountered in written texts and almost never in spoken texts. The most common concessive marker is the conjunction *although*, with the respective subordinate clause being the most common concessive gram-matical construction. In addition, *although* is one of the few markers that is semantically unambiguous, i.e., it can only introduce concession and not other relations. Therefore, it is a very convenient marker to be used in automatic extraction of concessive sentences from text, hence we will use it in our method-ology.

3   Methodology

In this section, we propose a way of extracting defeasible common-sense knowl-edge for a particular domain or topic using concessive markers. The vastly pre-vailing concessive construction in English is a subordinate clause with the con-junction *although*, which precedes the main sentence, e.g., *Although John studied hard for the exam, he failed it.* In a nutshell, our approach takes **the pattern P = "although X, Y"**, instantiates X with a particular value X_i (i.e., some property or category), queries a Web search engine with the first part of the construction *although X_i*, extracts full concessive sentences, and finds corresponding Y_is. The (X_i, Y_i) pairs are then used to generate defeasible statements *if X_i, then usually not Y_i.* Below is a step-by-step description of the methodology illustrated by an example that explores a personal category of gender (where the X_is are *male, female* etc.).

1. **Define a category of interest (select X_is)**
   In order to construct instances of the *although*-pattern, one needs to collect particular values for the chosen category. The selection can be done manually or using existing knowledge resources, e.g., WordNet or Wikipedia.

   For example, when interested in gender, one can search the term *gender* in WordNet[1] and collect related terms: *male, female, man, woman* etc.

2. **Build pattern instances P_is**
   Using the general pattern P and the chosen X_is, we construct pattern instances as follows: *although + he/she is (a) + X_i + , + he/she (is)*
   - *he/she* is used as a subject, since the gender category relates to persons;
   - an article *a* is optional and depends on the grammatical category of X_i;
   - the "*, he/she (is)*" part locks the beginning of the main clause; this guarantees that the main clause of the sentence refers to the same subject as the subordinate one, and we are more likely to get a direct opposition between the content of two clauses;
   - if we add *is* to the end of the instance, we are more likely to get another class or category as Y_i; without *is*, the main clause may be of arbitrary content, e.g., action, general description, event, etc.;
   - further variations of the main clause can include *he/she is not, does, does not* etc.

   In our example one possible pattern instance is *although she is a woman, she.*

3. **Query a search engine and crawl results**
   Pattern instances are used as exact queries for a search engine, so that the Web is treated as a text corpus. The search results are then crawled and the text snippets are collected.

   One of the utterances that was crawled for the example query is *Although she is a woman, she is fighting to have high degree education.*[2]

4. **Parse results and form (X_i, Y_i) pairs**
   Using state-of-the-art linguistic processing tools (e.g., Stanford CoreNLP[3]) we extract the Y_i part from the search result snippets:
   - we locate the main clause in the snippet (*clause identification*);
   - since snippets are restricted in size, a snippet may contain only a part of the main clause, in which case it is ignored altogether; this allows us not only to improve parsing, but also to filter out sentences that are too long

---

[1] http://wordnetweb.princeton.edu/perl/webwn?s=gender.
[2] Source: http://novrianfathi.blogspot.co.uk/.
[3] http://stanfordnlp.github.io/CoreNLP/.

and are hard to transform into a common-sense statement (e.g., *although she is a woman, she has control over the men in the bar because she is able to beat them at a ...*);

- we find the predicate of the clause and normalize it;
- if a sentence contains clauses preceding the *although*-clause (e.g., *She said that although she is ...*), they are removed;
- if the clause contains additional phrasal and parenthetical elements (*to tell the truth*), they are removed.

In our example, X_i is *woman* and Y_i is *fight to have high degree education*. Currently we leave the parsed segments as is, but in future, the Y_is can potentially be modified using synonyms (*higher education*), rephrasing (*want to have a degree*), generalization (*study*) etc.

5. **Negate Y_is**

   Argument negation is done to re-construct the defeasible assumption. It can be done using antonyms (*she is brave* vs. *she is fearful*), verb negation (*she is not brave*) or an introductory clause *it is not true that*.

   From our example, the defeasible assumption is: *if she is a woman, then she does not fight for high degree education.*

4   Proof of Concept Evaluation

Suppose we are interested in common-sense facts about the *Person* category. We can collect a set of X_i values manually, or we can as well address external knowledge resources. For example, Wikipedia has a rich network of categories, *People and self* being among the top 12 categories[4]. Subcategories can be a source of relevant aspects of the chosen category (e.g., gender, ethnicity, religion, occupation), as well as of the X_i values (e.g., male, female; Cherokee, Korean; Muslim, pantheist; diplomat, engineer). From these values pattern instances P_is can be constructed.

Let us consider the gender subcategory and the example query Q from Sect. 3: *although she is a woman, she*. We queried google.com with Q[5], collected snippets from the search results, filtered out certain snippets as discussed in step 4 of the methodology and saved the top 50 results. The list of extracted *although*-sentences can be found in the Appendix A. With the exception of several sentences that contain references to the context of the original web page (see the discussion section), other *although*-sentences can easily be converted into default implications:

(1) women are typically fearful (antonym of *fearless*), are not skilled at riding (verb negation), poor hunters (antonym to *excellent*), and bad warriors (antonym to *fine*), (44) women are good at cooking (verb negation), (32) women do not have male power in their work, etc.

The two main types of sentences are: those describing atypical behavior (*...she is fighting to have high degree education*), and those mentioning uncharacteristic

---

[4] https://en.wikipedia.org/wiki/Portal:Contents/Categories.
[5] Queried on 21.01.2016.

property (*...she is very sensible and smart*). While some sentences convey a narrower context than the others, they all are based on common accounts and stereotypes. Empirically, the shorter the second clause of the sentence, the more concise is the underlying statement and the easier it is to parse it automatically (compare: *she does not have kids* versus *she has never gone through the process of pregnancy and labor and delivery*).

Our methodology is based on utilizing the concessive pattern P = "*although X, Y*". P can be instantiated in numerous ways and proves to be quite general and versatile. While we demonstrated how it can be used to mine stereotypes about personal categories, we will now illustrate how the pattern instance "*although she is a woman, she*" can be: (a) narrowed down to target specific aspects of the category, and (b) generalized to more general categories. All examples are real-world and are queried using the specified pattern instances.

When narrowing down, we can mine particular common-sense statements that represent preferences, characteristics, actions, behavior by partially specifying the structure of Y_i:

- *...she likes/enjoys/prefers, ...she does not like* – preferences,
- *...she attends/works/does* – actions, activities,
- *...she speaks/plays/sews* – capabilities etc.

On the other hand, we can generalize pattern instances P_is by utilizing in the X_i the wildcard operator * that acts as a placeholder for any terms: "*although she is (a) *, she*". The resulting concessional statements reflect the most common stereotypes which can involve a female (but are not necessarily bound by female gender): professional (*Although she is a highly qualified graduate, she can't find work.*), personal (*Although she is a responsible adult she has a lot of kid in her and she seems to be having as much fun as the kids.*), religious (*Although she is a devout Christian, she observes the letter, but not the spirit, of the commandment "Honor thy father.*), related to hobbies (*Although she is a slow runner, she has completed 5 marathons!*) etc.

Finally, we can go beyond personal category and mine stereotypes and common-sense beliefs about pretty much any entity, however abstract or concrete:

- *although <u>London</u> is *, it*"
  Example: *Although London is an expensive city, it's also one that has lots of free attractions.*)
- *although <u>math</u> is *, it*
  Example: *Although math is challenging, it becomes easier with practice and a little bit of fun!*
- *although <u>inspiration</u> is *, it*
  Example: *Although inspiration is an amazing thing, it's often temporary and wears off long before our goals are accomplished.*
- *although <u>Pulp Fiction</u> is *, it*
  Example: *Although Pulp Fiction is full of violence, most of it isn't directly shown.*
  etc.

5   Related Approaches

To illustrate the specificity of defeasible common-sense knowledge and the contribution of this work, we will compare our methodology with three existing approaches for acquiring common-sense knowledge.

WebChild[12] is a project that automatically mines noun-adjective pairs from text, connecting concepts represented by nouns with their typical properties represented by adjectives via a set of predefined taxonomic and non-taxonomic relations (*hasShape, hasColor, hasTaste* etc.). The resulting triples are basic, common-sense facts, e.g., *apples are round.* The system has a major drawback compared to our approach: the set of statements is strictly limited by the semantic relations, whereas we are not confined by any particular item-property structure. When queried with the word "woman"[6], WebChild returns a number of basic statements: *woman type_of female,woman has_substance tissue* - as well as some more involved properties: *romantic, emotional, beautiful.* The latter can be viewed as common beliefs or stereotypes, but they do not go beyond category-property scheme, whereas our approach is able to retrieve much more versatile statements, e.g., women are good at cooking.

Verbosity[13] is a common-sense knowledge acquisition approach conducted in a form of a game: one player selects a concept and describes its typical properties without naming the concept itself, while the other player tries to guess the concept. The types of hints the first player can give are restricted to a predefined set of patterns with blanks to be filled in (*it is used for X, it is a kind of X*). Patterns considerably facilitate processing of the input sentences and enforce high precision of the acquired statements, but they lack expressivity. Our approach is generic and results in much more diverse statements. Another advantage of our approach is its scale: when Verbosity was played by over 250 people during one week, it managed to collect less than 8,000 statements. Using an automated approach, this amount of facts can be generated in a matter of minutes. Interestingly, the authors mention as reason for using gamification rather than simply querying the Web that common-sense statements are "too obvious" to be stated explicitly. As discussed earlier, our approach is precisely designed to overcome this issue.

NELL [7] is a large-scale project of common-sense knowledge harvesting, with a knowledge base of over 80 million statements collected from Web texts. NELL is very expressive and has a massive set of categories and relations, although it should be noted that the majority of statements are made on the level of instances (*ronnie_wood is a musician who is part of rolling_stones*). As in the case of WebChild, NELL is not tailored to mine defeasible knowledge, since the two items can only be linked together if they co-occur in text. Moreover, NELL is a system that is similar to freebase.com or dbpedia.org in a sense that it collects facts rather than beliefs, thus it targets other types of assertions. We can, on the other hand, complement NELL and other projects by applying our defeasible knowledge mining pipeline over their sets of categories, augmenting

---

[6] https://gate.d5.mpi-inf.mpg.de/webchild/.

their fact base with beliefs, stereotypes and popular opinions (and even common misconceptions).

6 Discussion and Future Work

A number of issues worthy of discussion correspond to particular steps of the proposed methodology. The methodology is a stepwise process, and each step of the pipeline can be customized; in particular, using a better web search crawler (more hits per query) or a better parser (more accurate clause identification and filtering) will result in better overall performance (more well-formed statements).

- **Building pattern instances**
  When a search engine is queried with a pattern *although X*, certain values for X tend to generate fewer results that the others. For example, *although she is a woman* returns approx. $325,000$ results, while *although she is an astronaut* returns only 3 results[7]. In order to maximize the number of concessive sentences containing the chosen value, one could expand the repository of patterns used for querying, both syntactic (*he is very weak for a footballer* producing a statement *football players are strong*) and semantic ones (*it is a very surprising fact that...*). For now, we stick to *although*-patterns, since they have a very high precision in terms of yielding concessive constructions. Expanding the number of patterns, on the other hand, would increase the recall.

- **Querying a search engine**
  Crawling turned out to be one of the implementation challenges of our approach. While Web data is the most comprehensive and useful text resource for the task at hand, the main access point to it is a search engine, and relying on a particular search engine means depending not only on its quality and the volume of indexed resources, but also on its data policies.

- **Parsing results and forming (X_i, Y_i) pairs**
  Converting *although*-sentences into concessive pairs is not a trivial task. While we can control the structure of X_is, Y_is can come in any shape and length, and some of them are harder to process than the others. Here are the examples of types of concessional sentences, ranging from "easier" to "harder" ones, together with the issues they contain and the potential remedies:
  - Y_i is short, easy to parse and to convert
    Example: *although she is a woman, she is brave*
  - Y_i contains a reference to a broader textual context
    Example: *although she is a woman, she is prepared to do it*
    The sentence can only be understood if the reference from *it* is resolved
    Solution: Such sentences can potentially be filtered out by using a *coreference resolution* module: when an unresolved item is encountered, the sentence is ignored.
  - Y_i is long, but can be split into several predicates
    Example: *although she is a woman, she is fearless, skilled at riding, an*

---

[7] As queried by google.com on Feb, 10th 2016.

excellent hunter, and a fine warrior
Solution: using syntactic parsing, we can divide the main clause into several predicates and corresponding subordinate words and negate them separately.

- Y_i is long and has a complicated inner structure
Example: *although she is a woman, she has only a woman's body and a woman's charm without a woman's heart*
It is not possible to shorten the sentence and get rid of some of its parts, because its meaning can be expressed only when the sentence is taken as one unit.
Solutions: there are two possible strategies with respect to such sentences. Either a more intricate parser procedures are implemented, or we filter out such sentences, focusing on snippets that can easily be parsed and transformed into defeasible assumptions. The latter take on common-sense knowledge mining seeks to, above all, maintain high precision rather than gain high recall and maximize the number of generated pattern instances.

- **Negating the second argument of the concessive relationship**
One of the most challenging steps in the suggested methodology is to transform an utterance of the form *although p, q* into a structured form *if p, then usually not q* and, in particular, to negate *q*. In multiple cases we need to go beyond antonyms and verb negation and to perform generalization in order to get meaningful statements. For example, *although he is German, he lives in France* implies that most German people live in Germany, but we cannot generate this statement by simply negating *France*.

Finally, there are several aspects of a general nature. Since languages are characterized by their flexibility and variability, the linguistic relation of concession does not impose strict constraints on the types of arguments it may use. Thus, defeasible beliefs and stereotypes may be as simple as comparing two facts (*X is German, but lives in France*), or as complex as involving several steps of reasoning and further background knowledge (*Although X loves Da Vinci, he did not enjoy Louvre*; $K_{background}$ =*Louvre is a museum; Louvre contains the Mona Lisa painting; Mona Lisa is painted by Da Vinci*). The more implicit reasoning steps are required to understand the sentence, the harder it is to automatically extract the default implication behind it. On the other end of the spectrum, there are sentences that exhibit an apparent contradiction: *Although the ending was a happy one, it was also a little sad*. The *happy-sad* opposition is actually independent of the context and thus is of little interest.

Some sentences extracted by our pipeline cannot be converted into common-sense facts, since they are very context-specific and do not rely upon common-sense knowledge. For example, a sentence *although she is a woman, she does not make any efforts to understand young Hazal's sentiments* uses the default implication that a woman would understand Hazal's sentiments, which is only shared between those who is aware of the broader context. The implication does not hold in general and cannot be viewed as a common-sense knowledge.

One last point to discuss is evaluation and quality insurance of the extracted knowledge. One needs to validate whether a generated default implication is indeed an instance of common-sense knowledge. Automatic evaluation against existing knowledge bases tends to be unreliable, since defeasible knowledge is not well-represented in the latter. A more feasible approach would be human evaluation, with several participants evaluating the same fact and with inter-annotator agreement being calculated for every item.

7   Conclusions

In this paper, we proposed a novel approach to extract common-sense knowledge from textual resources on the Web. Thereby, we overcome the sparsity of explicit occurrences of this type of knowledge by focusing on cases where the common-sense knowledge – being defeasible in nature – is violated. In such cases the inherent "contrariness" of two facts requires an explicit mention.

In linguistic terms, the violation of a common-sense assumption is typically expressed by concessional statements for which a variety of linguistic markers are known.

Given a domain of interest, we are able to systematically search the web for instances of concessive lexico-syntactic patterns and to extract utterances of the form *although p, q*, which can then be transformed into defeasible common-sense rules *if p, then usually not q*.

We gave an experimental proof of concept for our proposed methodology by extracting common beliefs and stereotypes about people, however, the suggested methodology can be adapted to various conceptual domains (e.g., organizations, events, artifacts) and types of common-sense information (e.g., typical actions, properties, relations).

The work is of exploratory nature: it serves as a proposal and first step toward accessing defeasible beliefs through the semantic relation of concession and its linguistic representation and paves the way for further research in common-sense knowledge acquisition and modelling.

A   Appendix

Although-sentences extracted from top 50 Google search result snippets for the query *although she is a woman, she* (incomplete sentences are ignored):
Although she is a woman, she...

1. ...is fearless, skilled at riding, an excellent hunter, and a fine warrior.
2. ...is strong and capable of keeping his secrets.
3. ...is not seen as one in the book.
4. ...is equally as capable of doing farmwork as the men are.
5. ...has some influence, and warns Krogstad to avoid offending his superiors.
6. ...displays serious proof of having "balls."?
7. ...is prepared to do it.

8. ...does not make any efforts to understand young Hazal's sentiments
9. ...can endure the march as well as any man.
10. ...is the muscle of the family.
11. ...has the physique of a man with broad shoulders.
12. ...has the heart of a king and that the invasion by the Spanish Armada is still "foul."
13. ...cannot bare working with women and this is reflected through her manners.
14. ...must demonstrate the "courage, ingenuity, and selflessness that is associated with Disney's male heroes".
15. ...is responsible for the tavern with her husband and she questions Falstaff without hesitation.
16. ...has nous within her.
17. ...has only a woman's body and a woman's charm without a woman's heart.
18. ...is determined to surpass men.
19. ...is similar in many ways to Jack LaLanne.
20. ...has brought science, enlightenment, and "masculine" rationality to the "female" Orient.
21. ...is the dominant one in her relationship and is known for all her accomplishments in "The Family".
22. ...is not seen as one.
23. ...believes it is only herself who can achieve her own fulfilment.
24. ...is more of a man than you.
25. ...is prepared to die.
26. ...holds the same power and authority as all the men who have ruled before her.
27. ...was born a boy.
28. ...is brave.
29. ...has never gone through the process of pregnancy and labor and delivery.
30. ...has more of an ambitious like a man compared to Duncan.
31. ...is very sensible and smart.
32. ...has a male power in her work.
33. ...fights to revive her ruined homeland
34. ...is hardly worth considering to be a sex object.
35. ...still "manned up".
36. ...acts like man, so we can consider her a male.
37. ...will rule alone.
38. ...is equal to the occasion.
39. ...believes in chivalry.
40. ...has lofty aspirations.
41. ...acts much like a warrior, fighting alongside her Thenns like any other knight.
42. ...can do thing, which helps safe her family while her father loses the power as a family protector.
43. ...doesn't use the typical features of women's writing
44. ...isn't good at cooking.
45. ...fights like a man.

46. ...has much confidence
47. ...is fighting to have high degree education
48. ...has not lost the wonder and playfulness of a child.
49. ...often can be more dependable and confident than men.
50. ...is still old yet mysterious and attractive to men.

References

1. Davis, E.: Representations of commonsense knowledge. Morgan Kaufmann, Burlington (2014)
2. Fensel, D.: Ontologies - a silver bullet for knowledge management and electronic commerce. Springer, Heidelberg (2001)
3. Grice, H.P.: Logic and conversation. In: Cole, P., Morgan, J. (eds.) Syntax and Semantics. Speech Acts, vol. 3, pp. 41–58. Academic Press, New York (1975)
4. Izutsu, M.N.: Contrast, concessive, and corrective: Toward a comprehensive study of opposition relations. J. Pragmatics 40(4), 646–675 (2008)
5. Koons, R.: Defeasible reasoning. In: Zalta, E.N. (ed.) The Stanford Encyclopedia of Philosophy (2014). http://plato.stanford.edu/archives/spr2014/entries/reasoning-defeasible/, spring 2014 edn
6. Lam, X.N., Vu, T., Le, T.D., Duong, A.D.: Addressing cold-start problem in recommendation systems. In: Proceedings of the 2nd International Conference on Ubiquitous Information Management and Communication, pp. 208–211. ACM (2008)
7. Mitchell, T., Cohen, W., Hruschka, E., Talukdar, P., Betteridge, J., Carlson, A., Dalvi, B., Gardner, M., Kisiel, B., Krishnamurthy, J., Lao, N., Mazaitis, K., Mohamed, T., Nakashole, N., Platanios, E., Ritter, A., Samadi, M., Settles, B., Wang, R., Wijaya, D., Gupta, A., Chen, X., Saparov, A., Greaves, M., Welling, J.: Never-ending learning. In: Proceedings of the Twenty-Ninth AAAI Conference on Artificial Intelligence (AAAI-15), pp. 2302–2310 (2015)
8. Moses, Y., Shoham, Y.: Belief as defeasible knowledge. Artif. intell. 64(2), 299–321 (1993)
9. Nute, D.: Defeasible logic. In: Gabbay, D., Hogger, C., Robinson, J.A. (eds.) Handbook of Logic in Artificial Intelligence and Logic Programming. Nonmonotonic Reasoning and Uncertain Reasoning, vol. 3, pp. 355–395. Oxford University Press, New York (1994)
10. Robaldo, L., Miltsakaki, E.: Corpus-driven semantics of concession: Where do expectations come from? Dialogue & Discourse 5(1), 1–36 (2014)
11. Taboada, M., de los Ángeles Gómez-González, M.: Discourse markers and coherence relations: Comparison across markers, languages and modalities. Linguist. Hum. Sci. 6(1), 17–41 (2012)
12. Tandon, N., de Melo, G., Suchanek, F., Weikum, G.: Webchild: harvesting and organizing commonsense knowledge from the web. In: Proceedings of the 7th ACM International Conference on Web Search and Data Mining, pp. 523–532. ACM (2014)
13. Von Ahn, L., Kedia, M., Blum, M.: Verbosity: a game for collecting common-sense facts. In: Proceedings of the SIGCHI Conference on Human Factors in Computing Systems, pp. 75–78. ACM (2006)
14. Winter, Y., Rimon, M.: Contrast and implication in natural language. J. Semant. 11(4), 365–406 (1994)

Visualizing Ontologies – A Literature Survey

Arash Saghafi[(✉)]

Sauder School of Business, University of British Columbia, Vancouver, Canada
arash.saghafi@sauder.ubc.ca

Abstract. Information, as a representation of the "real world", is required to faithfully demonstrate the relevant aspects of the application domain. To describe the structure of a domain, various fields have employed ontological models. Visualization of the said ontologies can improve tasks such as understanding of implicit knowledge as well as information alignment. The present work provides a survey on the methodologies proposed for visualizing ontologies in the literature, and also performs a statistical synthesis (i.e. meta-analysis) to quantitatively review some of the empirical studies that focused on the impact of visualization enhancement.

Keywords: Ontology · Visualization · Survey · Meta-analysis

1 Introduction

Ontology is a branch of philosophy[1] that deals with the order and structure of reality [31]. Considering that information systems are representations of applications, practitioners as well as researchers in information and computer sciences have used ontologies as guidance to describe the order and structure of domains in order to develop more faithful representations of reality [25, 26]. Domain ontology is defined as a set of concepts, the relationship between concepts, what can happen, and what can exist - the axioms [31]. In computer science, ontology is a formalization of reality that can be implemented in a computer system [24].

Diverse fields such as biomedical informatics, systems engineering, and semantic web[2] have developed ontologies to represent the semantic meta-data within their fields. One of the largest ontologies available is the ontology of the DBpedia project[3], which is a cross-domain ontology with over 4.2 million resources (things) in the ontology.

Visualizations of ontologies have been proposed in prior research [17, 22]. Visualization in general is created to augment human capabilities in performing a task [21]. Some of the tasks that can benefit from visualizing ontologies could be implicit knowledge identification in a domain [1], integration of data sources [11, 23], and understanding a domain in general [22]). The objective of this paper is twofold: (1) Survey the existing literature focusing on visualization of domain ontologies. And (2) synthesize the data from similar empirical experiments evaluating different ontological visualizations (i.e. meta-analysis).

---

[1] The term "ontology" was first coined in 1613 in works by Rudolf Göckel and Jacob Lorhard [24].
[2] http://protegewiki.stanford.edu/wiki/Protege_Ontology_Library. Accessed on 08/12/2015.
[3] http://dbpedia.org/services-resources/ontology. Accessed on 14/03/2016.

© Springer International Publishing Switzerland 2016
O. Haemmerlé et al. (Eds.): ICCS 2016, LNAI 9717, pp. 204–221, 2016.
DOI: 10.1007/978-3-319-40985-6_16

2 Ontology Language and Baseline Representation

The standard language for developing ontologies, according to the World Wide Web Consortium (W3C) is the Web Ontology Language or OWL (http://www.w3.org/2001/sw/wiki/OWL). The most widely used tool for creating and modifying ontologies is an open source program called Protégé [10], – developed and maintained at Stanford University[4]. The ontologies created in Protégé are represented as indented trees (or lists), similar to the structure of files in Windows Explorer (or Finder in Mac OSX). In OWL, the class called "Thing" is the root of the class hierarchy of the ontology; this class (i.e. "Thing") includes all instances in the relevant universe (http://www.w3.org/2001/sw/wiki/OWL). The other classes in the domain ontology are defined based on the properties that they posses. All the classes in the ontology are assumed to be subclasses of "Thing". Figure 1 illustrates a hierarchy of a sample ontology that models menu items in a pizzeria, while Fig. 2 shows the properties in this domain.

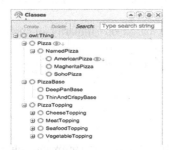

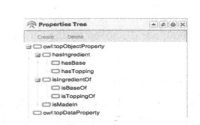

Fig. 1. Classes of the pizza ontology in Protégé   **Fig. 2.** Properties in the pizza ontology

3 Survey of Visualization Methods in the Literature

As part of research for the presented work, three literature surveys on visualizing ontologies were identified. The first, and most influential[5], survey was done by Katifori et al. [15]. They presented methods tailored for visualizing ontologies, as well as other techniques (for different contexts) that could also be used to display ontologies. They also performed an empirical user study [14], by testing four different visualizations – all of which were plug-ins in Protégé. Two of the methods were based on indented-lists (hierarchies), and the other two were node-link based. The experimental task was investigating evolution of entities in an ontology over time. They found that users of Protégé class browser (indented lists) perform this task with higher accuracy.

Another survey was done by Lanzenberger et al. [17]. Their purpose was to identify techniques that could be used for the task of ontology alignment, or in other words, reconciling the meta-data from various sources (i.e. interoperability). Their focus was

---

[4] http://protege.stanford.edu. Accessed on 08/12/2015.
[5] Cited 340 times based on Google Scholar data, as of 08/12/2015.

mostly on graph-based visualization tools, whether 2D or 3D. At the end, they concluded that a compelling method that utilizes the screen real estate appropriately, while being intuitive to users, is yet to be developed.

The third survey Granitzer et al. [11] also focused on the task of interoperability. They studied various Protégé plug-ins that enabled reconciliation of two or more distinct ontologies. Similar to the previous survey, after discussing the strengths and weaknesses of the current approaches, they stated that the field lacks a comprehensive technique for ontology alignment. They described the appropriate approach as a semi-automatic technique that would utilize human judgment and at the same time, handle complex and evolving ontologies.

For the purpose of current work, 21 papers were identified that had investigated methods to visualize ontologies. One of the selection criteria for these papers was their exclusion in the surveys done by [11, 15, 17]. The publication date of the collected papers was in the period of 2003–2014, with the majority being published after 2011. Out of this pool of 21 papers, 11 of them had done some sort of empirical user evaluation. These evaluations ranged from lab experiments to interviews and protocol analyses.

The review in this work groups the studies based on similarity of methods and presents them in the following sections. In each group of similar visualizations, the methods are discussed by the chronological order of publication.

3.1   Graph-Based Methods

3.1.1   Spring Embedded/Force-Directed

Mutton and Golbeck [22] used a spring embedding algorithm to draw the ontology as a graph. This visualization is similar to the force-directed placement visualization [3, 21], in which each class in the ontology is considered a node, while the links of the graph represent the relationship between classes in the ontology – that is whether a class is subclass of another, sibling class, or completely disjoint from the other class (declared by a closure axiom in OWL).

This algorithm considers visualizing an ontology similar to simulating a force system: the nodes act as charges particles, thus the repulsive forces between the nodes in the graph imposes a layout where similar nodes end up being placed closer to each other. This visualization is most useful for the task of identifying similar concepts in a domain ontology, as similar concepts will be found in a cluster within the spring embedded graph. Figure 3 illustrates such visualization of an ontology.

Fig. 3. Spring embedded graph visualization of ontologies [22]

Vercruysse et al. [30] proposed a similar visualization (i.e. force-directed graphs) for ontologies. Their data source was biomedical ontologies available on the Ontology Lookup Service (OLS) database. The task that they investigated in their paper was just browsing ontological graphs by the users (as uses of ontologies in biomedical research are diverse, and this paper did not want to limit the scope to one specific task). A force-field graph visualization grants the users flexibility in the way they view the ontologies (and sub-ontologies), as the graphs can be reorganized smoothly, and enables moving the concepts "towards more optimal positions" [30, p. 4] in the canvas, hence improving the exploration of bio-ontologies.

In a more recent paper, Fua et al. [8] empirically evaluated spring-embedded graph representations versus indented lists using eye-tracking method with 36 subjects. Their justification for their evaluation was that "a lack of scientific evaluations of existing ontology visualization techniques could be potentially damning to the advancement of this field as a whole, as we may fail to recognize and adopt good designs, or to identify and reject bad practices" (p. 1).

Fua et al. [8] found that indented lists are more efficient in tasks involving information search – defined by the authors as tasks "where [subjects] only need to sample a small amount of objects to complete" (p. 7), while information processing tasks – interpretation of information that are measured by duration of eye fixation – are done more efficiently using graph based representations. They also measured accuracy (i.e. error rate) and completion time. Completion time was faster for indented lists, however accuracy was not significantly different between the two visualization methods.

3.1.2 Clinical Outcome Search Space (COSS)

As mentioned earlier, one of the tasks that could be facilitated by visualizing ontologies, is identification of implicit knowledge within a domain. Andronis et al. [1], used ontologies in biomedical domain for the task of drug repurposing (DR). They posited that instead of researching and developing new drugs, biomedical practitioners could utilize ontologies of existing drugs, and test different hypotheses regarding their effects. The premise of this study is that complete and comprehensive ontologies of drugs are available to researchers. They developed a mining technique for drug repurposing called Clinical Outcome Search Space (COSS), which acts as a semantic reasoner; COSS analyses ontologies and establishes similarities between various drugs and the symptoms they treat. Using graph-based techniques with colour coding, they visualize the drugs that could be repurposed. It should be noted that COSS is not an ontology visualization technique, but a visualization of inferences made on ontologies. Figure 4 shows the relations detected between two genes within an ontology.

3.1.3 Graph Formation Based on Information Richness

Motta et al. [20] suggest a graph-based approach, in which the effectiveness of 'sense-making' of the topology of an ontology will be improved. In their method, they place the concept with highest information richness – measured by density of properties and taxonomic relationships of a concept – at the centre of the graph. Using a 'middle-out' approach, their proposed tool (called KC-Viz) organizes the other nodes around the centre (i.e. the node with highest information richness). They evaluated this approach with an empirical experiment: 21 subjects were asked to answer questions

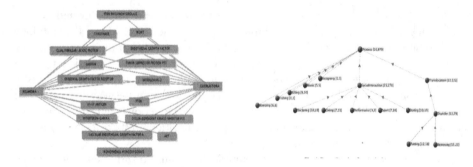

Fig. 4. COSS graph-based visualization [1]   **Fig. 5.** KC-Viz [20]

(i.e. problem-solving task) using KC-Viz and also OWLViz (the default visualization tool of Protégé). They found out that efficiency of performance was higher for KCViz (i.e. time taken to complete the task was shorter on average). From a subjective evaluation point of view, subjects were asked to report their satisfaction levels as well. [20] showed that KC-Viz led to higher satisfaction among users. Figure 5 shows an example of an ontology in KC-Viz.

3.1.4   Wheel-Graph

To visualize content within an ontology (based on user-defined filters), Tscherrig et al. [29] proposed wheel-graph visualization. Rather than visualizing the whole ontology, they assumed that users are searching for particular information. From the ontology, one can infer what relationships exist, and what relationships can happen (i.e. axioms); thus, based on the axioms, this paper tries to visualize the relationship between search criteria. The authors developed a prototype called Memoria-Mea project. Memoria-Mea creates wheel graphs for each search criterion, and based on the relationship between different criteria, visually structures results of each parameter within a layer.

As an example, they explored an ontology related to travelling. Some of the relevant concepts (classes) in this ontology are: "country", "people", "geographical location", and "activity". The relationships between these classes (i.e. the axioms) are: "people" visit "countries", "people" go to a "geographical location" within a "country", and "people" do some type of "activity" at a "geographical location". Based on these assumptions, when a user searches for the activities done by people who visited Switzerland, Memoria-Mea will generate a layered wheel graph visualization of the results similar to Fig. 6.

The authors also did a usability study with six subjects. All subjects found the required information using Memoria-Mea (i.e. successful performance of a task). [29] also evaluated[6] users' subjective perceptions of the visualization; overall subjects found the interface to be easy to use, and "not frustrating".

---

[6] Descriptive statistics were not provided for this study.

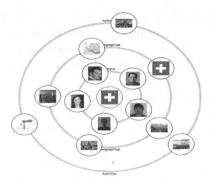

Fig. 6. Wheel graph visualization of content in an ontology [29]

3.2 Multi-method Visualization Techniques

3.2.1 Graphs and Space-Filling Blocks

In order to improve understandability of OWL ontologies, Jurcik and Sochor [13] introduced a plug-in for Protégé, called Knoocks. In this visualization method, each node in the graph is a space-filling block that represents class hierarchies. A block is designated for each subclass of OWL:Thing. The subclasses of the said class will be represented in a hierarchy. In the example from Fig. 7, "Destination" is a subclass of OWL:Thing. Some of the subclasses of "Destination" are "Country", "Town", and "City".

This visualization method is particularly useful for displaying an overview of the whole ontology, in order to achieve a general understandability. As can be seen in Fig. 8, the relationships between blocks are visualized as edges between the blocks. Different colours represent different types of relationships. For example, pink is used to model the relationship titled "Travels to", and connects "Passenger" and "Destination" blocks.

3.2.2 Lists and Linked Histograms

As a design study paper, Streicher and Roller [28] presented a visualization to display semantic search results. The application domain was image interpretation, for tasks such as pollution detection, cartography, ice layer monitoring and surveillance. These

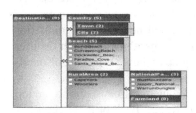

Fig. 7. A block in Knoocks [13]

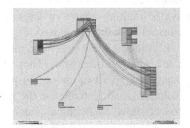

Fig. 8. Overview of an ontology in Knoocks [13] (Color figure online)

tasks are done on data gathered by Synthetic Aperture Radar (SAR) sensors, and then stored based on "a domain ontology that encompasses concepts of the specific field of work" [28, p. 51]. Their motivations for using ontologies were the facilities that ontologies provide in interoperability (of different data sources), as well as the advantages in performing semantic searches.

The proposed visualization is composed of two views: first view presents a list of the facilities for which SAR data has been gathered. The ranking of this list could be based on the frequency of classes of data for each facility. The second view, which is linked to each of the facilities in the list, presents histograms of signal strengths of different classes of sensory data. The visualization proposed in this study, also allows for inverse lookup of locations based on a special class of sensory data. More specifically, the first view displays classes of sensory data based on location, while the second mechanism, lists locations in which a certain class of sensory data are present. Figures 9 and 10 represent the two features respectively.

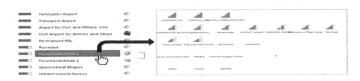

Fig. 9. Displaying histograms based on a location on the list [28]

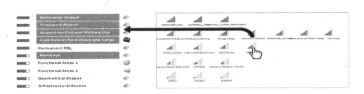

Fig. 10. Displaying locations that feature a certain class of sensory data [28]

3.2.3 Hierarchical Connected Circles (Radial Layout), Indented Trees, Node-Link Diagrams

In order to support seven high-level tasks of overview, zoom, filter, details-on-demand, relate, history, and extract, Kuhar and Podgorelec [16] proposed a visualization tool that provided multiple views of large and complex ontologies (i.e. high scale). The implementation that they proposed, displays hierarchical connected circles (to provide overview), indented trees (to relate different concepts), and node-link diagrams (for filtering and details-on-demand). They also designed a toolbar through which the user could change the speed of animation for a dynamic graph (showing history of ontology's evolution), and also choose the level of semantic zoom.

Figure 11 shows hierarchical connected circles; the outermost circle shows the top level classes in the ontology (direct children of OWL:Thing). Each of these classes is coded with a colour (different hues). Subclasses of the aforementioned classes are visualized within inner circles, coded with different levels of saturation of the parent's colour. The relationship between classes are also modelled as links, connecting different segments of circles to each other.

Figure 12 shows a small section of the circle, in which the user has zoomed in to see the information related to the class of "Professor" in more detail. Figure 13 shows the toolbox available for filtering, zooming, and animation. Figure 14 displays the indented tree and node-link diagrams of the same ontology.

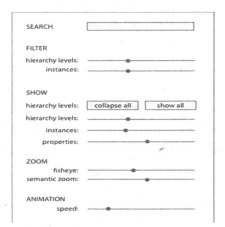

Fig. 11. Visualizing ontology as hierarchical connected circles [16] (Color figure online)

Fig. 12. Semantic zoom on one of the classes in the ontology [16] (Color figure online)

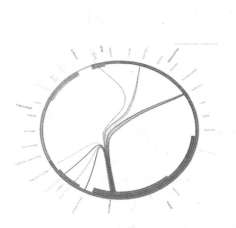

Fig. 13. Filtering, zooming, and animation [16]

Fig. 14. Indented tree and node link visualization of the university ontology [16] (Color figure online)

Another (somehow similar) tool was developed by Ma et al. [19] as part of a design study that visualized geological time-scale data. They also used radial layout (or hierarchical circles) with colour coding, however, they placed the superclasses in the inner circles, while the subclasses were represented in the outer circles (as opposed to the visualization proposed by [3]). Figure 15 shows a snapshot of this tool.

[19] also did a user study with 19 participants. They asked the users to navigate through the visualization and at the end answer a usability survey. Subjects' average scores were between "useful" and "very useful".

3.2.4   Hyperbolic and Radial Trees

Another technique proposed to facilitate comprehension of the overall view of the ontology (i.e. concepts and their relationships) is implemented in OntoViewer – a tool developed by da Silva et al. [6]. This tool facilitates visual exploration of large ontologies by employing three integrated views: "a hyperbolic tree for representing the ontology hierarchy; a classic tree view for showing ontology entities, and an augmented radial tree for displaying relationships between classes" [6, p. 93]. The tree view is similar to the indented list that is provided by Protégé (Fig. 1). The 2D hyperbolic tree provides focus + context features and reduces the cognitive load of users when interacting with a large ontology. The hyperbolic trees that this paper suggests is similar to force-directed graphs (in Fig. 3) that others had also proposed [22, 30].

The most interesting visualization in OntoViewer is the 2.5D radial tree – which is to some extent similar to hierarchical connected circles (proposed by [3]). The class hierarchy is modelled in the radial tree: the class under focus will be at the centre, and all of its subclasses will orbit (or circle) around it. The relationships between classes (i.e. mutual properties between two classes) are "represented as curved lines in space (thus yielding 2.5D), connecting the related classes without interfering with the display of the hierarchical structure" [6, p. 93]. The tool also allows for viewing the ontology

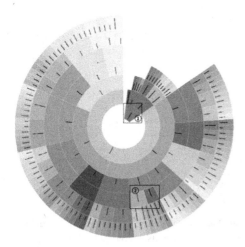
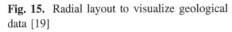

Fig. 15. Radial layout to visualize geological data [19]

Fig. 16. 2.5D radial tree visualization of an ontology [6]

"by choosing to display one or more relationships at the same time or hiding them, choosing which levels of the tree are to be shown or hidden, performing rotations around the axes X, Y and Z, zoom and pan, i.e. providing full 3D navigation" (p. 93). Figure 16 shows the 2.5 radial tree visualization.

The authors posit that providing multiple and coordinated views in OntoViewer will help in pattern recognition and revealing hidden relationships in large ontologies.

In a later paper, same authors [7] discussed the possibility of visualizing ontologies using OntoViewer at intensional as well as extensional levels. Intensional level representation deals with classes and relationships between classes (i.e. data schema), while extensional level represents individual observations or instances of classes with specific property values. The authors point out that intensional level is more important from the point of view of knowledge engineers as they may "want to visualize different aspects due to specific demands that arise in certain stages of development, for example, checking the range of an object property" (p. 2). Extensional representations, on the other hand, are claimed to be "more interesting from the point of view of professionals that maintain knowledge databases [since] it seems necessary to have views of the synthetic instances distribution allowing to see how attribute values are distributed and to perform quick visual queries about instances, observing trend in values" [7, p. 2]. The features of OntoViewer introduced earlier enables all the needs of knowledge engineers for viewing intensional level representation of ontologies. As for extensional representation, the authors built upon the OntoViewer tool, and added another view by employing overview + detail methods using an icicle tree. Figure 17 shows an icicle tree representing the instances of the "Worker" class. This visualization is similar to the space-filling blocks in Knoocks [13].

The OntoViewer tool also provides a hybrid view (intensional and extensional), as depicted in Fig. 18. The authors also evaluated their tool by interviewing small group of experts – all of whom found the tool "effective".

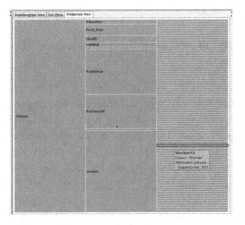
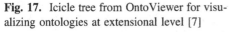
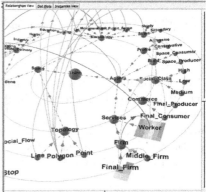

Fig. 17. Icicle tree from OntoViewer for visualizing ontologies at extensional level [7]

Fig. 18. Intensional and extensional representation of an ontology in OntoViewer [7]

3.2.5   Treemap, Hiearchies, and Radial Layout

Another multi-method tool is Onto-VisMod by Garcia-Penalvo et al. [9] that incorporates treemaps to represent large ontologies (which "uses the whole available space in the dimensional plane"), hierarchical trees to analyse the taxonomy of concepts, and radial layouts to represent the global coupling of an ontology.

The treemaps and hierarchies can be transposed onto each other to give a taxonomy of the domain, while making use of "two-dimensionally squared maps, where the lower levels are represented as internal squares located inside the higher level maps" (p. 11472). Figure 19 captures a snapshot of this view.

Onto-VisMod uses radial layouts to visualize the relations among classes in the ontology. On one side of the circle is a list of classes, the other side list of properties. Properties that are used in definition of a class are linked to it. User has the ability to focus on a particular class: when a class is selected, properties that define it will be highlighted, and the links become coloured. This visualization method is different from hierarchical connected circles [16], as the hierarchy of classes is not modelled here, and in addition, the user would see the list of all properties in the relevant universe (i.e. application domain). Figure 20 shows the radial layout view in Onto-VisMod.

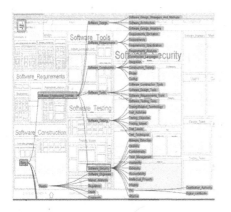

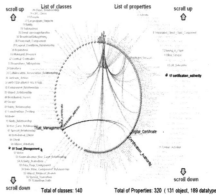

Fig. 19. Treemap and hiearchies within an ontology [9]

Fig. 20. Radial layouts in Onto-VisMod [9]

The authors did an empirical evaluation of this tool with 21 subjects. The experimental task involved navigation through the ontology and creating a new class. The same task was also done by the users with Protégé. The results of the experiment showed no statistically significant difference in performance (i.e. completion of the task). However, users' satisfaction scores with Onto-VisMod were higher (than their satisfaction with Protégé).

3.3   Euler Diagrams

ConceptViz is a tool developed by Burton et al. [4] that employs Euler diagrams to visualize topological properties of ontologies and provide an overview of the semantic information that an ontology conveys. This approach might be limited in scale, as it only shows the relationships between a few concepts (classes) at a time. However, this method could visually represent set inclusion when a curve is contained by another – reflecting a subsumption relationship. In short, this method is most useful for providing an abstract description of the ontology. Figure 21 shows an example in ConceptVis. Visualizing three concepts in the "Pizza" ontology, it shows that the class of thin and crispy pizzas is a subset of specialty pizzas. At the same time, one can infer that deep pan pizzas and thin and crispy pizzas are disjoint concepts, however, some deep pan pizzas are considered specialties of the pizzeria.

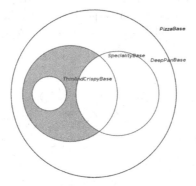

Fig. 21. Euler diagram visualization of the pizza ontology [4]

Using Euler diagram visualization of ontologies, Howse et al. [12] did a case study, and visualized Semantic Sensor Networks (SSN) ontology. SSN is an ontology in the philosophical sense (i.e. structure and order of reality in a broad sense) rather than representation of a single domain. SSN was developed by the World Wide Web Consortium (W3C); in this ontology the world is made of objects that sense and make observations. Using Euler diagrams, [12] demonstrated how they could merge simple axioms into more complex axioms to describe the world.

4   Methods to Model Ontologies

Significant focus has been put on modelling ontologies in the literature. Here, three of those studies will be briefly discussed. Silva-Lopez et al. [27], introduced a graphical notation – called Onto Design Graphics (ODG) – based on UML components. They posit that it could be a standard notation for ontology engineering research; it is easy to learn and could be an efficient method in design and integration of ontologies. The contribution of this work was establishing a formalized mapping between ontological concepts and UML constructs.

Console et al. [5] proposed another modelling grammar – named Graphol - as an alternative to UML-based visualization of ontologies. In Graphol grammar, symbolic constructs are assigned to each ontological concept; for example, rectangle for a concept nodes (i.e. class), circle for attributes, and hexagon for individuals (i.e. instances) in the ontology. They performed an empirical evaluation of Graphol (by comparing with UML-based visualizations); the average correctness scores in comprehension tasks were not statistically different between the two groups (Graphol vs. UML-based), however, users reported Graphol to be easier to use than other methods - This user study was basically a comparison between two graph-based models.

Visual Web Ontology Language (VOWL), is a noteworthy method, which is developed by Lohmann et al. [18]. The constructs of this grammar are primitive shapes and a colour scheme (shown in Figs. 22 and 23 respectively). A user study with six expert users was conducted to evaluate VOWL. The experimental tasks were comprehension questions, and overall, "participants could solve most of the tasks correctly (84 %)" [18, p. 11]. This was, however, a qualitative user study in order to verify the general ideas of VOWL, and also receive feedback from users in order to enhance VOWL for non-expert (novice) ontology users.

Primitive	Application
○	classes
═	properties
▷ ►	property directions
▭	datatypes, property labels
- - - -	special classes and properties
text number symbol	labels, cardinalities

Name	Color	Application
General		classes, object properties, disjoints
Rdf		elements of RDF and RDF Schema
External		external classes and properties
Deprecated		deprecated classes and properties
Datatype		datatypes, literals
Datatype property		datatype properties
Highlighting		highlighted elements

Fig. 22. VOWL graphical primitives [18]

Fig. 23. VOWL colour scheme [18] (Color figure online)

Figure 24 shows a visualized ontology created with VOWL.

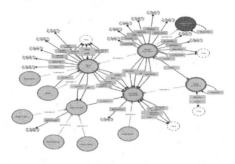

Fig. 24. An ontology in VOWL [18]

5   Meta-analysis

Among the papers that were reviewed in this survey, 11 of them had done a user study. Some of them were qualitative user studies with limited number of participants. However, four of these studies are relatively similar as they study one form of visualization enhancement and compare it with the baseline view of Protégé.

To perform a quantitative review of these studies, a statistical synthesis (i.e. meta-analysis) of their findings was conducted; such analysis enables reflection upon the findings of the past researchers [2].

A meta-analysis could be conceptualized as either a fixed-effects model or random-effects [2]. Fixed-effects model assumes that all the studies in the meta-analysis are identical and they share a common effect size. Any variation that exists between the findings of different studies in the pool would be due to sampling error. "Put another way, all factors that could influence the effect size are the same in all the studies" [2, p. 63].

The random-effects model, on the other hand, incorporates a group of studies in a meta-analysis, assuming that they have "enough in common that it makes sense to synthesize the information, but there is generally no reason to assume that they are identical" [2, p. 69]. The variation in different studies is attributed to sampling error, as well as Random Effects Variable (i.e. the variation between studies, such as the chosen variables for the study, or the experimental methods).

The four studies that were analysed here had different independent and dependent variables, yet they all had focused on the impact of visualization on some measure of user performance. Due to the fact that studies in the pool are not identical, random-effects model was chosen for the purpose of this analysis.

5.1   Data Coding

The statistics available from these studies were in the form of means and standard deviations. Therefore, to perform the synthesis (i.e. meta-analysis), the findings of the studies were converted to Cohen's d (which is the standardized mean difference between two groups [2]). Since the sample sizes were different in each study, Cohen's d-values needed to be unbiased; this was done by assigning weights to the d-values according to their respective standard errors [2]. Positive effect sizes mean that visualization enhancement led to an improvement in performance (e.g. faster completion time, or higher comprehension score) – and vice versa for negative effect sizes. Table 1 represents the data in the analysis.

5.2   Analysis and Discussion of Results

The average unbiased Cohen's d of this analysis is 0.014, with the 95 % confidence interval of -1.21 to 1.22. This means that an alternative visualization (compared to indented lists in Protégé) can improve a measure of performance by 0.014 standard deviations from the mean (i.e. the average performance achieved by users of indented lists) – which is a very weak effect (almost none). 95 % of the time, the impact of the

Table 1. Meta-analysis data

Reference	Independent variable	Dependent variable	Sample size	Effect size (unbiased Cohen's d)
[14]	TGViz (Node-link) vs. Protégé	Accuracy	23	−0.49
[20]	KCViz (Node-link) vs. Protégé	Completion time	21	1.01
		Usability score	21	0.26
[9]	OWL-VisMod (tree + radial layout) vs. Protégé	Accuracy	21	−0.14
		Usability score	21	0.79
[8]	Node-link vs. Protégé	Completion time	36	−1.32
		Accuracy	36	0

alternative visualization falls in the reported credibility interval. In other words, 95 % of the time, an alternative visualization method could reduce performance effectiveness by −1.21 standard deviations or improve it by 1.22 standard deviations. Grouping the effect sizes based on dependent variables provides additional insights, as seen in Table 2. Accuracy and completion time for users of the baseline representation (i.e. Protégé) is superior to the performance of users when they use alternative visualization methods. However, subjects' perceptions (usability score and satisfaction) will be higher when they have access to alternative visualization methods. The usability score has 95 % confidence interval of 0.20 to 0.84, meaning that users will find (mostly graph-based) alternatives more appealing.

Table 2. Grouping the variables of the meta-analysis

Dependent variable	Average unbiased Cohen's d	No. of reported effect sizes	95 % credibility interval
Accuracy	−0.21	3	−0.45 to 0.12
Completion time	−0.16	2	−2.90 to 2.56
Usability score	0.53	2	0.20 to 0.84

As possible justification for these findings, one could (tentatively) hypothesize that subjects' perceptions of an alternative visualization method might be influenced by its novelty and differentiation compared to the baseline representation (i.e. indented lists in Protégé); thus, positive impressions lead to higher usability scores for alternative visualization methods. However, as the meta-analysis shows, objective measures - accuracy in particular - are usually higher when users access the indented list representations. This could be due to nature of ontological data, in which hierarchies are inherent to them. Indented lists incorporate the hierarchy, and also are familiar representations to users of computer systems (as used in Windows Explorer, Mac OSX Finder).

6 Summary and Conclusions

The present work gathered 21 different studies that had focused on visualization of ontological data. This pool included a mixture of method and design study papers. Each paper was described, and categorized based on its respective visualization method.

Moreover, a meta-analysis was done on four papers that even though they were not identical, they had enough in common to justify synthesizing their findings. These studies had compared alternative visualization methods (three of them studied graph-based visualizations) with the baseline representation of ontologies, which is indented list in Protégé. According to the meta-analysis, the only factor that significantly improves is human users' perceptions of usability - which is a subjective measure. Objective measures of accuracy and task completion time did not show a significant change (neither for the better nor worse). But this might be due to the limited number of studies included in the meta-analysis.

The present study calls for more effective visualization methods that incorporate the hierarchical nature of data (which are inherent in ontologies); the new methods need to focus on improving objective measures of user performance such as accuracy of answers and task completion time.

Acknowledgment. Gratitude to my mentor Yair Wand for his influential role on my career thus far, and special thanks to Tamara Munzner for her comments and feedback as instructor of the graduate course in Information Visualization that catalyzed this work.

References

1. Andronis, C., Sharma, A., Virvilis, V., Deftereos, S., Persidis, A.: Literature mining, ontologies and information visualization for drug repurposing. Brief. Bioinform. **12**, 357–368 (2011)
2. Borenstein, M., Hedges, L.V., Higgins, J.P., Rothstein, H.R.: Introduction to Meta-analysis. Wiley, Chichester (2009)
3. Brandes, U.: Drawing on physical analogies. In: Kaufmann, M., Wagner, D. (eds.) Drawing Graphs. LNCS, vol. 2025, pp. 71–86. Springer, Heidelberg (2001)
4. Burton, J., Stapleton, G., Howse, J., Chapman, P.: Visualizing concepts with Euler diagrams. In: Dwyer, T., Purchase, H., Delaney, A. (eds.) Diagrams 2014. LNCS, vol. 8578, pp. 54–56. Springer, Heidelberg (2014)
5. Console, M., Lembo, D., Santarelli, V., Savo, D.F.: Graphol: ontology representation through diagrams (2014)
6. da Silva, I., Santucci, G., del Sasso Freitas, C.: Ontology visualization: one size does not fit all. In: International Workshop on Visual Analytics, EuroVA 2012, pp. 91–95. The Eurographics Association (2012)

7. da Silva, I.C.S., Freitas, C.M.D.S., Santucci, G.: An integrated approach for evaluating the visualization of intensional and extensional levels of ontologies. In: Proceedings of the 2012 BELIV Workshop: Beyond Time and Errors-Novel Evaluation Methods for Visualization, p. 2. ACM (2012)
8. Fua, B., Noyb, N.F., Storeya, M.A.: Eye Tracking the User Experience-An Evaluation of Ontology Visualization Techniques (2014)
9. García-Peñalvo, F.J., Colomo-Palacios, R., García, J., Therón, R.: Towards an ontology modeling tool. A validation in software engineering scenarios. Expert Syst. Appl. 39(13), 11468–11478 (2012)
10. Gašević, D., Djuric, D., Devedžic, V.: Model Driven Engineering and Ontology Development, 2nd edn. Springer, Heidelberg (2009)
11. Granitzer, M., Sabol, V., Onn, K.W., Lukose, D., Tochtermann, K.: Ontology alignment—a survey with focus on visually supported semi-automatic techniques. Future Internet 2(3), 238–258 (2010)
12. Howse, J., Stapleton, G., Taylor, K., Chapman, P.: Visualizing ontologies: a case study. In: Aroyo, L., Welty, C., Alani, H., Taylor, J., Bernstein, A., Kagal, L., Noy, N., Blomqvist, E. (eds.) ISWC 2011, Part I. LNCS, vol. 7031, pp. 257–272. Springer, Heidelberg (2011)
13. Jurčík, A., Sochor, J.: Knoocks-ontology visualization plug-in for protege. In: Proceedings of CESCG 2012: The 16th Central European Seminar on Computer Graphics (2012)
14. Katifori, A., Vassilakis, C., Lepouras, G., Torou, E., Halatsis, C.: Visualization method effectiveness in ontology-based information retrieval tasks involving entity evolution (2006)
15. Katifori, A., Halatsis, C., Lepouras, G., Vassilakis, C., Giannopoulou, E.: Ontology visualization methods—a survey. ACM Comput. Surv. (CSUR) 39(4), 10 (2007)
16. Kuhar, S., Podgorelec, V.: Ontology visualization for domain experts: a new solution. In: 2012 16th International Conference on Information Visualisation (IV), pp. 363–369. IEEE (2012)
17. Lanzenberger, M., Sampson, J., Rester, M.: Ontology visualization: tools and techniques for visual representation of semi-structured meta-data. J. UCS 16(7), 1036–1054 (2010)
18. Lohmann, S., Negru, S., Haag, F., Ertl, T.: VOWL 2: user-oriented visualization of ontologies. In: Janowicz, K., Schlobach, S., Lambrix, P., Hyvönen, E. (eds.) EKAW 2014. LNCS, vol. 8876, pp. 266–281. Springer, Heidelberg (2014)
19. Ma, X., Carranza, E.J.M., Wu, C., van der Meer, F.D.: Ontology-aided annotation, visualization, and generalization of geological time-scale information from online geological map services. Comput. Geosci. 40, 107–119 (2012)
20. Motta, E., Mulholland, P., Peroni, S., d'Aquin, M., Gomez-Perez, J.M., Mendez, V., Zablith, F.: A novel approach to visualizing and navigating ontologies. In: Aroyo, L., Welty, C., Alani, H., Taylor, J., Bernstein, A., Kagal, L., Noy, N., Blomqvist, E. (eds.) ISWC 2011, Part I. LNCS, vol. 7031, pp. 470–486. Springer, Heidelberg (2011)
21. Munzner, T.: Visualization Analysis and Design. A K Peters Visualization Series. CRC Press, New York (2014)
22. Mutton, P., Golbeck, J.: Visualization of semantic metadata and ontologies. In: Proceedings of Seventh International Conference on Information Visualization, IV 2003, pp. 300–305. IEEE (2003)
23. Parsons, J., Wand, Y.: Attribute-based semantic reconciliation of multiple data sources. In: Spaccapietra, S., March, S., Aberer, K. (eds.) Journal on Data Semantics I. LNCS, vol. 2800, pp. 21–47. Springer, Heidelberg (2003)
24. Øhrstrøm, P., Andersen, J., Schärfe, H.: What has happened to ontology. In: Dau, F., Mugnier, M.-L., Stumme, G. (eds.) ICCS 2005. LNCS, vol. 3596, pp. 425–438. Springer, Heidelberg (2005)

25. Recker, J., Rosemann, M., Green, P.F., Indulska, M.: Do ontological deficiencies in modeling grammars matter? MIS Q. **35**(1), 57–79 (2011)
26. Shanks, G., Tansley, E., Nuredini, J., Tobin, D., Weber, R.: Representing part-whole relations in conceptual modeling: an empirical evaluation. MIS Q. **32**, 553–573 (2008)
27. Silva-López, R.B., Silva-López, M., Méndez-Gurrola, I.I., Bravo, M.: Onto Design Graphics (ODG): a graphical notation to standardize ontology design. In: Gelbukh, A., Espinoza, F.C., Galicia-Haro, S.N. (eds.) MICAI 2014, Part I. LNCS, vol. 8856, pp. 443–452. Springer, Heidelberg (2014)
28. Streicher, A., Roller, W.: Semantically driven presentation of context-relevant learning material. Int. Proc. Econ. Dev. Res. **37**, 50–55 (2012)
29. Tscherrig, J., Carrino, F., Sokhn, M., Mugellini, E., Khaled, O.A.: Ontology based scope visualisation. In: 2012 IEEE/WIC/ACM International Conferences on Web Intelligence and Intelligent Agent Technology (WI-IAT), vol. 3, pp. 210–214. IEEE (2012)
30. Vercruysse, S., Venkatesan, A., Kuiper, M.: OLSVis: an animated, interactive visual browser for bio-ontologies. BMC Bioinform. **13**(1), 116 (2012)
31. Wand, Y., Weber, R.: Research commentary: information systems and conceptual modeling —a research agenda. Inf. Syst. Res. **13**(4), 363–376 (2002)

Key Discovery for Numerical Data: Application to Oenological Practices

Danai Symeonidou[1], Isabelle Sanchez[1], Madalina Croitoru[2]($\boxtimes$), Pascal Neveu[1], Nathalie Pernelle[3], Fatiha Saïs[3], Aurelie Roland-Vialaret[4], Patrice Buche[5], Aunur-Rofiq Muljarto[1], and Remi Schneider[4]

[1] INRA, MISTEA Joint Research Unit, UMR729, 34060 Montpellier, France
[2] University of Montpellier, Montpellier, France
croitoru@lirmm.fr
[3] LRI (CNRS UMR8623 & Université Paris Sud),
Université Paris Saclay, Orsay, France
[4] NYSEOS, Montpellier, France
[5] INRA, IATE Joint Research Unit, UMR1208, 34060 Montpellier, France

Abstract. The key discovery problem has been recently investigated for symbolical RDF data and tested on large datasets such as DBpedia and YAGO. The advantage of such methods is that they allow the automatic extraction of combinations of properties that uniquely identify every resource in a dataset (i.e., ontological rules). However, none of the existing approaches is able to treat real world numerical data. In this paper we propose a novel approach that allows to handle numerical RDF datasets for key discovery. We test the significance of our approach on the context of an oenological application and consider a wine dataset that represents the different chemical based flavourings. Discovering keys in this context contributes in the investigation of complementary flavors that allow to distinguish various wine sorts amongst themselves.

1 Introduction

Nowadays, more and more data are produced in the context of life science. Analysing and understanding such data is a task of great importance. A series of classical methods is often exploited for this analysis: [7,8,11]. Nevertheless, these approaches are not always explicative enough especially when data are described using many properties. To address more efficiently this problem, we propose to use a *key* discovery method. Key discovery has been initially introduced in the context of relational databases and then in the Semantic Web in order to better understand the data. A key represents a set of properties that uniquely identifies every resource in the data.

Some recent approaches [5,6,12,15,16] have been developed to discover automatically keys from RDF datasets. In this paper we use SAKey [16] that takes into account the OWA in key semantics [4]. Indeed, discovering keys in RDF datasets without taking into account RDF data specificities (such as erroneous data and duplicates) may lead to discovery of irrelevant keys or false negatives.

O. Haemmerlé et al. (Eds.): ICCS 2016, LNAI 9717, pp. 222–236, 2016.
DOI: 10.1007/978-3-319-40985-6_17

Furthermore, certain sets of properties that are not keys but share a small number of shared values, can be useful for data linking or data cleaning. These sets of properties are particularly needed when a there are only few valid keys.

While key discovery has already been tested on large datasets with the above mentioned characteristics (such as DBpedia [2] and YAGO [3]) in SAKey [16], many real world RDF datasets contain numerical data describing results of scientific experiments. The state of the art in key discovery in RDF datasets does not handle experimentally issued numerical data due to the following challenges that *motivate our contribution*:

- First, numerical data are too precise to enable the discovery of relevant keys. Indeed, numerical data as such hold no semantic meaning (especially when handled as Strings).
- Second, the size of the available data is in general not big enough to discover only relevant keys.

Different measures have been used in the literature to evaluate the quality of the keys, like support and discriminability [6,16]. However, in the context of numerical data describing scientific experiments, additional quality measures have to be defined.

In this paper we propose *a novel approach that allows to handle numerical RDF datasets for key discovery*. To deal with the numerical data, a preprocessing step is applied to convert numerical data into symbolic data. To do so, we use statistical methods to group numerical data in classes. By grouping the data into classes we exploit the closeness of values of data. The use of quantiles ensures that the values are distributed in equal-size groups. The numbers of groups were defined in order to have significant probability mass. Based on the obtained symbolic classes we apply SAKey to discover keys that are valid in this preprocessed data. Finally, two new quality measures are defined and used to evaluate the discovered keys. The first method is based on the property value distribution in the dataset and the second method is based on correlations that can be found between key properties.

In order to demonstrate the practical use of our method we test this approach on a dataset that describes wines obtained from the Pilotype project[1]. The dataset contains a set of numerical values regarding different chemical components that give the flavour of wines. In this application setting, the discovered keys can be used to discover flavour complementarity, unknown from the experts, that allow to distinguish various wine sorts amongst themselves. We have also described the dataset from a statistical point of view using principal component analysis (PCA) on the raw data (without quantile classification). This allowed us to have a more global view of the dataset and to see the way the properties (i.e., the different flavours) are correlated amongst each other. We have then

---

[1] http://www.qualimediterranee.fr/projets-et-produits/consulter/les-projets/theme-1-agriculture-competitive-et-durable/das2-tic-chaine-alimentaire/theme-1-devel-opper-une-agriculture-competitive-et-durable/das-2-contribution-des-tic-a-la-chaine-alimentaire-en-amont/pilotype.

validated the keys obtained with domain experts and discussed their interest with respect to the statistical analysis.

The contribution of this paper is *three fold*:

- From a *methodological point of view* we extend upon the state of the art in two directions. We provide the first approach in the literature dealing with key discovery numerical data issued from experiments. We introduce novel quality evaluation criteria for such keys and discuss them in context of existing work.
- From an *application point of view* our results are extremely important as we use key discovery on (1) a domain where the interpretation of keys can only be done by domain experts and (2) for a dataset that prevents the discovery of such keys manually by the expert. This calls for both *(i)* statistical analysis to quantitatively validate the different keys found and *(ii)* a qualitative analysis to validate the interest of keys for the application.
- From an *interdisciplinary point of view* this paper successfully demonstrates the added value of putting together results issued from Computer Science, Statistics and Oenology.

This paper is organised as follows. In Sect. 2 we introduce the methodology proposed for the key evaluation. Then, in Sect. 3 we show how our method is applied in the case of a Wine dataset and we present the experimental results. In Sect. 4, we provide an analysis of the keys discovered and finally in Sect. 5 we discuss about the related works existing in this domain and we propose future works.

2 Numerical Key Discovery Methodology

A key is a set of properties that uniquely identifies every resource in a data knowledge base. In the context of Semantic Web, keys are defined for a specific class. Keys are used in several tasks such as data linking but can also be very useful to understand the data.

In this work we employ an automatic key discovery method, as the task of defining keys is very difficult for domain experts. Experts may define erroneous keys. Moreover, the more the data are described by many properties, the harder it becomes for a human expert to define all the possible keys. Please note that in experimental setting data experts may not be aware of the existence of keys. Furthermore, it has been shown in [12] that keys automatically discovered are relevant and can even be more significant than keys defined by experts.

Our numerical key discovery approach aims to discover keys in numerical experimental data. The approach is performed in several steps shown in Fig. 1. In the preprocessing step, numerical data are transformed to obtain a symbolic interpretation. Then we apply SAKey to discover a set of minimal keys.

In what follows, we first present the preprocessing step, then the key discovery approach used in this work is described. Finally, several measures that can be used to evaluate the quality of the discovered keys are proposed.

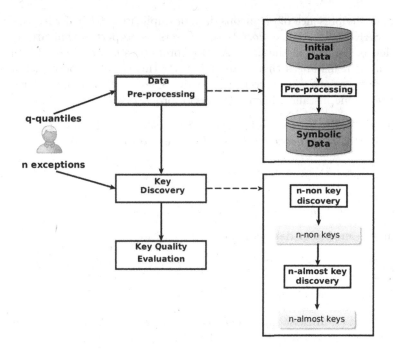

Fig. 1. Method steps followed in this paper

2.1   Data Pre-processing

As previously explained, if we treat the data as purely numerical we get a lot of keys (since the numerical data is treated as simple Strings). Therefore, we need a way to cluster the data. Quantiles are cutpoints dividing a set of observations into equal sized groups. q-Quantiles are values that partition a finite set of values into q subsets of (nearly) equal sizes. By using quantiles, we reduce the number of values and potentially decrease the number of naive keys, so we can now take advantage of knowledge of how the properties that might work together form keys. Different strategies for computing the quantiles can be found in [11].

Intuitively, accepting only few groups may lead real keys to be not lost. On the contrary, allowing many groups can lead to the discovery of keys that are not valid in the real life. In general, more the data are grouped in few groups/many groups, more the number of properties that are involved in a key increases/decreases respectively. Therefore, choosing the appropriate size of groups can play a very significant role in the obtained results.

2.2   Key Discovery

We use SAKey [16] to discover automatically keys in the preprocessed data. SAKey is an approach that is able to discover keys in very large RDF datasets even under the presence of erroneous data or duplicates. To deal with the errors in data, SAKey discovers *almost keys*, i.e., sets of properties that are not keys due to few exceptions in the data. A set of properties is a n-almost key if there exist at most n instances that share values for this set of properties. Formally, the set of exceptions E_P corresponds to the set of instances that share values with at least one instance, for a given set of properties P.

Definition 1 (Exception set). *Let c be a class ($c \in C$) and P be a set of properties ($P \subseteq \mathcal{P}$). The exception set E_P is defined as:*

$$E_P = \{X \mid \exists Y (X \neq Y) \wedge c(X) \wedge c(Y) \wedge (\bigwedge_{p \in P} \exists U p(X,U) \wedge p(Y,U))\}$$

where X and Y are instances of the class c.

For example, in the D1 dataset showed in Fig. 2 we have:

$$E_{\{d1:producer\}} = \{w1, w3\}$$

Using the exception set E_P, a n-almost keyis defined in [16] as follows:

Definition 2 (n-almost key). *Let c be a class ($c \in C$), P be a set of properties ($P \subseteq \mathcal{P}$) and n an integer. P is a n-almost key for c if $|E_P| \leq n$.*

This means that a set of properties is considered as a n-almost key, if there exist from 1 to n exceptions in the dataset.

By definition, if a set of properties P is a n-almost key, every superset of P is also a n-almost key. Therefore, this approach discovers only minimal n-almost keys, i.e., n-almost keys that do not contain subsets of properties being n-almost keys for n fixed.

To illustrate this approach, let us introduce an example. Figure 2 contains descriptions of wines. Each wine can be described by the region it produced in, the name of the producer and finally its color.

In this example, the property $d1$:*region* is not a key for the class *Wine*. Indeed, there exist two wines, $w1$ and $w2$, produced in *Bordeaux* and similarly, $w3$ and $w4$ in *Languedoc*. Considering each wine that shares the region of production with other wines as an exception, there exist 4 exceptions for the property $d1$:*region*. Thus, the property $d1$:*region* is a 4-almost key in this example since it contains at most 4 exceptions. Obviously, this property does not refer to a meaningful key since it contains the maximum number of exceptions (i.e., the number of exceptions is equal to the number of instances). Similarly, the property $d1$:*color* is a 2-almost key. Regarding composite keys, the composite key $\{d1$:*region*, $d1$:*producer*$\}$ is a 0-almost key. Similarly, $\{d1$:*producer*, $d1$:*color*$\}$ is a 0-almost key.

The problem of discovering the complete set of keys automatically from the data is #P-hard [9]. Therefore, using a simplistic way to discover keys would not be appropriate. To validate if a set of properties is a n-almost key for a class c in a dataset D, a naive approach would scan all the instances of a class

---

Dataset D1:
d1:Wine(w1), d1:*region*(w1," *Bordeaux*"), d1:*producer*(w1," *Dupont*"),
d1:*color*(w1," *White*"),

d1:Wine(w2), d1:*region*(w2," *Bordeaux*"), d1:*producer*(w2," *Baudin*"),
d1:*color*(w2," *Rose*"),

d1:Wine(w3), d1:*region*(w3," *Languedoc*"), d1:*producer*(w3," *Dupont*")
d1:*color*(w3," *Red*"),

d1:Wine(w4), d1:*region*(w4," *Languedoc*"), d1:*producer*(w4," *Faure*"),
d1:*color*(w4," *Red*").

Fig. 2. Example of wine RDF data

c to verify if at most n instances share values for these properties. Even in the cases where a class is described by few properties, the number of candidate n-almost keys can be huge. For example, let us consider a class c that is described by 60 properties. In order to discover all the n-almost keys that are composed of at most 5 properties, the number of candidate n-almost keys that should be checked would be more than 6 millions. Therefore, efficient ways to discover keys are necessary.

An efficient way to obtain n-almost keys, as already proposed in [12,14,16], is to discover first all the sets of properties that are not n-almost keys and use them to derive the n-almost keys. Indeed, to show that a set of properties is not a n-almost key, it is sufficient to find only $(n+1)$ instances that share values for this set. The sets that are not n-almost keys, are defined first in SAKey as n-*non keys*. A n-non key is defined as follows:

Definition 3 (n-non key). *Let c be a class ($c \in C$), P be a set of properties ($P \subseteq P$) and n an integer. P is a n-non key for c if $|E_P| \geq n$.*

For example, the set of properties $\{d1:region, d1:color\}$ is a 2-non key (i.e., there exist at least 2 wines sharing regions and colors).

SAKey discovers maximal n-non keys, i.e., n-non keys that are not subsets of other n-non keys for a fixed n. Indeed, as shown in [14], minimal keys can be derived from maximal non keys. Once the discovery of n-non keys is completed, the approach proceeds to the next step which is the derivation of n-almost keys from the set of n-non keys. In SAKey, an efficient strategy for the derivation of n-almost keys has been introduced.

2.3   Key Quality Measures

To evaluate the quality of the discovered keys, different quality measures are proposed. Since the number of discovered keys can be numerous, strategies that allow the selection of the most significant ones are needed. Depending on the characteristics of the dataset one or more of the following quality measures can be chosen.

Key Support. To begin with, the support, a classical measure used in data mining, can be exploited. The support, denoted as S_k, represents the ratio of number of instances described by the set of properties involved in a key k to number of instances described in the data. Intuitively, a key is significant when it is applicable to an important part of the data. Nevertheless, when all the instances are described by the same set of properties, the support cannot be exploited to compare the quality of different keys.

Key Exceptions. A second measure is the number of exceptions $|Ep|$ presented in Sect. 2.2 [16]. Even if keys with exceptions can be interesting, when the number of exceptions increases significantly, the quality of the discovered keys decreases.

Key Size. Keys composed of numerous properties can be difficult to evaluate. Therefore, we tend to select simple keys composed of few properties. This criteria is also taken into account in the next measure.

Key Probability. While discovering keys from the data, sets of properties that are not real keys may be discovered. More precisely, the property value distribution can be taken into account to evaluate the probability of a discovered key to have at least two instances that share values.

Given a set of properties, more it is probable to have at least two instances sharing all values for this set of properties, more this set of properties is considered significant when discovered as key [1].

For example, in a wine dataset describing 100 distinct wines, containing among others the properties $wineName$ and $YearProduction$. If we have 20 distinct wine names and 5 distinct years of production the probability that every couple of these properties is unique is very low. Therefore, if $wineName, YearProduction$ is discovered to be a key, we consider that its quality is very high.

We compute the probability using the following formula:

$$Pr_k = 1 - e^{-\frac{n(n-1)}{2p}}$$

where n is the number of instances described in the data, $p = \Pi_{i=0}^{j} m_i$ with j representing the number of properties participating in the key k and m_i is the number of distinct values of each property participating in the key.

In practice, we can set a threshold that acts as a probability lower bound that allows us to decide whether to consider a key or not.

Property Correlation and PCA. The more correlated the properties are the less informative a key is. Indeed, in such cases, its properties give intuitively the same kind of information [10]. PCA (Principal Component Analysis) is a multivariate procedure used to reduce the dimensionality of a dataset while retaining as much information as possible. PCA provides a description of the dataset by projecting the data onto new reduced dimensions that are linear combinations of the properties of the dataset. It can be used to graphically study the property correlation between properties with respect of the reduced dimensions. It also allows to visualise the individuals (the instances of the dataset and their respective values for the properties) on the reduced dimensions. However, sometimes, if the number of reduced dimensions is higher than a given threshold (still lower than the initial number of properties) the results could be difficult to interpret for a human expert.

Regarding key discovery, when the number of properties is to high, we consider that the correlation of the pairwise properties that make up a key is an indicator of how much the properties depend on each other. Please note that a correlation takes a value between -1 and 1.

3   Oenological Data Key Discovery

This section describes the practical results obtained for the method described above on a real dataset. The section is structured as follows. First, in Sect. 3.1 we describe the dataset used in the paper. We present in Sect. 3.2 the numerical experimental results: the preprocessing step of the dataset, the obtained keys and their quality measures.

3.1   Wine Dataset Description

Wine aroma data investigated in this paper have been obtained during a 4 years (2011–2014) research project called Pilotype. The aim of the project was to investigate the influence *(i)* of grape water stress on aroma precursors contents in berry and *(ii)* to study relationship between aromatic potential content in grapes and aroma profiling of corresponding wines. For this purpose, 3 different grape varieties (Chardonnay, Merlot and Grenache) from plots located in south of France (Languedoc-Roussilon region) were harvested then vinified according to three different winemaking processes. Enological, polyphenolic and aromatic analyses were performed along the winemaking process on grape, must and wine.

The data used here represent the different chemical based flavourings of wine. There are three datasets corresponding to three years of measurements: 2012, 2013, 2014. Each year measures the chemical flavouring of 63, 59 and respectively 44 wine instances. In each of the datasets used (2012, 2013, 2014), the wines are described using 19 flavourings (see column Analyzed molecules in Fig. 3).

In 2013 and 2014, each grape variety was collected from at least 3 different plots in order to have triplicate conditions. Three different processes dedicated to white (Chardonnay), rosé (Grenache) and Red (Merlot) wines were used to obtain experimental wines at pilot scale (1 hL) and under standardized conditions (pilot technical cellar).

Experimental wines were analysed by classical methods (GC-MS/MS and LC-MS/MS) in order to quantify important aroma compounds as described in Fig. 3. All these aroma compounds were quantified by Stable Isotope Dilution

Chemical families	Concentration levels in wine	Analyzed molecules	Analysis methodology
Thiols	ppt	3MH = 3 mercaptohexanol	LC-MS/MS
		3MHA = 3 mercaptohexylacetate	
Esters	ppm	2PHEN= 2-phenylethanol	GC-MS/MS
		AH= hexyl acetate	
		AI= isoamyl acetate	
		ABPE= phenethyl acetate	
		DE= ethyl decanoate	
		HE= ethyl hexanoate	
		OE= ethyl octanoate	
		BE= ethyl butyrate	
		2HPE= Ethyl lactate	
		3HBE= Ethyl 3-hydroxybutyrate	
		2MBE= Ethyl 2-methylbutyrate	
		2MPE= Ethyl isobutyrate	
		2HICE= Ethyl leucate	
C13-noriprenoïds	Ppb	BDAM= beta-damascenone	GC-MS/MS
		BION= beta-ionone	
PDMS	Ppb	Dimethylsulfide potential= S-methylmethione + others compounds	GC-MS/MS
GSH	ppm	Glutathione	LC-MS/MS

Fig. 3. Wine flavourings in Pilotype dataset

Assay (SIDA) to ensure high quality results in terms of accuracy and reproducibility.

Here, we are interested in automatically emerging any complementarity between flavours. Since, aroma compounds vary depending on the chosen year, the complementarities are studied year per year. In this context complementarity is understood in the sense that it allows for the flavours to combine in order to discriminate a fermentation method or not.

3.2   Experiments

The data quantify the flavour potential of different wine varieties (expressed as instances in the RDF file). We took measurements for several different flavours thus we have a lot of flavour properties. We are interested to see if there are flavours that either go well together or are complementary. This means that when certain flavours are formed, the other flavours are not and vice-versa. Our approach will be able to extract such information from the numerical data. If a set of properties is a key that means that the flavours they represent are complementary.

Data Pre-processing. The data pre-processing step has been performed using R [13], a programming language and software environment for statistical computing and graphics. In this experiment, we grouped the data in quantiles using the Gumbel strategy, considering the modal position [11]. This strategy is commonly used in the R language. We grouped the datasets using 5-quantiles,

10-quantiles and finally 12-quantiles. The size of quantiles has been chosen taking into account the size of instances in the datasets. Choosing a big number of groups, may lead to not significant probability mass. In our case, each group always contains between 8 % and 20 % of values for a given property.

When setting the size of quantiles to less than 5, no keys were obtained. Similarly, since the datasets are describing few instances, setting a very big size of groups leads to the discovery of insignificant keys. Therefore, empirically, we selected 3 different sizes of quantiles to investigate the effect of the size in the results. Finally, the size of quantiles affects the probability score of a key.

Key Discovery. In Table 1 we show the number of discovered keys for the three datasets corresponding to the years 2012, 2013, 2014. The keys are shown for the datasets split in 5-quantiles, 10-quantiles and finally 12-quantiles.

We have discovered many minimal keys. Thus, we have used the key size quality measure to reduce the number of keys that can be shown to the expert. More precisely, for each year and each quantile only the keys that have the minimal size have been selected (see the values in bold in Table 1). In total we showed 104 keys to the expert.

Table 1. Number of keys per year and per quantiles in Pilotype

Nb of properties/quantiles	2012			2013			2014		
	3	4	5	3	4	5	3	4	5
5	0	0	0	0	**2**	72	0	0	**32**
10	0	**23**	472	**3**	305	1348	**4**	454	471
12	0	**1**	149	**24**	461	705	**13**	415	664

Key Quality Measures. The expert has validated 18 keys among the 104 showed keys (i.e., 17.3 %). In Table 2 we show for each validated key the obtained values for support, probability and size.

A set of 11 validated keys have key probability greater than 40 %. This result shows that the probability measure allows to select a significant part of the relevant keys: the smaller the probability of a set of properties being a key, the more of interest the key can be.

In the validated keys the support varies from 63 % to 100 % and the key probability from 17 % to 94 %. The fact that the expert selected a key with not a full support shows that the support as such is not a meaningful indicator of the quality of a key. Similarly we thought that the lower the number of properties in the key, the more informative the key is. However, the expert did not always follow this postulate and the best informative keys were selected as the keys of 5 properties.

To further our study, a qualitative evaluation is also available for the first three keys - the keys of size 5. They were considered by the expert as the most

Table 2. Quality measures for expert validated discovered keys in Pilotype

	Year	Quantiles	Support	Probability	Size
[3MHA, BDAM, GSH, 2MPE, 3MH]	2014	5	73 %	26 %	5
[3MHA, GSH, OE, 2HICE, 3MH]	2014	5	73 %	26 %	5
[3MHA, AI, PDMS, 2MPE, 3MH]	2014	5	100 %	26 %	5
[3MHA, BE, PDMS, 2MPE, 3MH]	2014	5	100 %	26 %	5
[BDAM, OE, PDMS, 3MH]	2012	10	100 %	17 %	4
[GSH, OE, PDMS, 2PHEN]	2012	10	100 %	17 %	4
[AI, BDAM, 2HICE, 3MH]	2012	10	100 %	17 %	4
[3MHA, BDAM, GSH, 2MPE]	2013	5	63 %	94 %	4
[3MHA, BDAM, GSH]	2013	12	63 %	64 %	3
[AH, BDAM, GSH]	2013	12	63 %	64 %	3
[BE, 2HICE, 3MH]	2013	12	100 %	64 %	3
[BDAM, GSH, 3MH]	2013	12	63 %	63 %	3
[GSH, PDMS, 3HBE]	2013	10	63 %	83 %	3
[BDAM, GSH, 3MH]	2013	10	63 %	83 %	3
[GSH, PDMS, 3HBE]	2014	10	73 %	63 %	3
[PDMS, 3HBE, 3MH]	2014	12	100 %	44 %	3
[3MHA, GSH, PDMS]	2014	12	73 %	44 %	3
[BE, GSH, 3MH]	2014	12	73 %	44 %	3

informative keys as each property in the key corresponds to a class of aromas in wine (i.e., same bio-genetic origin). These aromas and their pertinence as keys are better explained in Sect. 4.

This experiment has shown that the probability of keys is not monotonic with respect to their size. Moreover, we noticed that the probability alone is not sufficient to select all the relevant keys: the keys that were found to be the most interesting are the ones with a relatively small (but not the smallest, nor the biggest) probability. Therefore, other quality criteria should be defined such as correlation presented in Sect. 4.

Please note that the probabilities can serve as a way to help the expert against his /her own confirmation biases. For instance, the first key has an extremely small probability and even if the expert invariably can find a justification, from a data analysis point of view the key is artificial.

As a conclusion, on top the quantitative evaluation, a qualitative evaluation is highly needed. One reason for this is the complexity of the domain and of the data which is not easily analysed by simple numerical measures. Therefore, in the next section we will also analyse the results from a statistical point of view and put our results in context of oenological practices with a domain expert.

4   Evaluation

In this section we present the analysis of our results from a practical perspective. We have worked in close contact with one of the wine experts involved in the project Pilotype and the main results explained here are drawn after discussion with the domain expert.

4.1   Statistical Key Analysis

The dataset was analyzed using bivariate and multivariate methods. Correlations of Pearson (linearity assumption) and Spearman (without assumption) were calculated pairwise [7]. A principal component analysis was used to get an global overview of the data [8].

We consider three of the discovered keys extracted from Table 2. The first key analysed is of size three: (BE, 2HICE, 3MH). The Pearson correlation of the pairwise properties of this key vary from 0 to 0.51. The second key of size four is (AI, BDAM, 2HICE, 3MH) varies from 0.05 to 0.42. Last, the key (3MHA, BE, PDMS, 2MPE and 3MH) varies from 0.02 and 0.49. Please note that a correlation of 0.5 is the highest correlation between these properties. Similarly, the Spearman correlation of (BE, 2HICE, 3MH) varied from 0.05 to 0.55. This confirms our hypothesis that because the properties are not significantly correlated their combination functions well as an interesting key.

We now provide the second statistical analysis that aims to multivariate analyse the data. We consider here the 2012 dataset. The PCA on our dataset on the two first principle components (Dimension 1 and 2) can explain 62 % of the variability. We can see that individuals are projected into the two first axis following the oenological modality.

The information we can extract from the PCA analysis on the first and second dimension is as follows. GSH and PDMS (present in the 6th key shown in Table 2) are badly explained by the PCA detailed in Fig. 4. 2PHEN is little explained by first axis and not much explained by the second axis. In general, aromatic properties are drowned within other properties and we cannot interpret them. This allows us to show the limits of the PCA analysis and its complementarity to the analysis of the key pertinence.

4.2   Oenological Key Significance

Please note the description of wine flavourings in Fig. 3. Here we discuss the qualitative value of the first three keys (the keys of size 5) found the most pertinent by the domain expert. These keys are composed of several aromas that are complementary to each other. Therefore they allow to better identify a wine. More precisely 3MHA and 3MH give some information on the thiol biogenesis. The higher the presence of thiol, the more the grapefruit and passion fruit like aroma is present in the wine. Such aromas are very appreciated by consumers. The PDMS gives an idea of the ageing potential of wine. The older the wine, the more DMS is released. The DMS plays an important role in the fruity taste

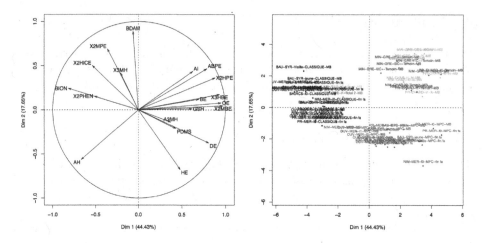

Fig. 4. PCA of 2012 dataset

of the wine. BE and AH are fermenting indicators and GSH gives an idea of the level of wine oxidation. Last 2MPE gives an indication of the fermentation level of the biogenesis.

We observe that most of the obtained keys contain a class of aroma per element of the key (property). This characteristic allows the expert to define wine aromatic profiles.

5   Discussion

In a context of life science, it is important to be able to simultaneously analyse large datasets with multiple properties on large population with the aim of a comprehensive understanding of the studied system. To this end, different clas-sical methods are available. These approaches are used for exploratory purposes in order to give more insights or descriptions into biological studies: Principal Component Analysis (global overview), Multiple Factor Analysis, Canonical Cor-relation Analysis or Partial Least Squares Regression (relationships of different types of properties in a non-supervised or supervised context). Sparse versions of these methods are then used in order to select relevant properties on data. This allows to explain most of the variance/covariance data structure using linear combinations of the original properties.

The great advantage of these geometrical methods is the possibility of repre-senting the data in reduced dimensions (usually the three first components) for a global overview. Yet, to be restricted to these first components may seem too much simplified, especially since these planar visualizations often suffer of a lack of biological interpretability and require statistical background skills. The appli-cation of these kinds of methods can also be limited by the size of the dataset (particularly for CCA) and the noisy and multicolinearity characteristics of the properties (PCA, CCA, MFA). Furthermore in the case of highly dimensional case, these methods are extremely cumbersome to use.

In this paper we presented a method that allows key discovery on oenology experimental numerical data rendered symbolic by means of quantiles. We have evaluated our method practically and tested it with domain experts.

The research avenues opened by this work are numerous. First the analysis of the correlation between probabilities and the keys discovery should be taken at a larger scale and, if possible, on synthetic datasets that will allow the better analyse it. Second, we could envisage an interface with the domain expert that will facilitate the understanding of keys. This method could be generalized for association rules and dependencies allowing to discover hidden rules within the numerical data. Last, using the Pearson correlation allowed us to see that certain properties are strongly correlated. Such attributes are important to be known as they might not bring complementarity when mining for key discovery. Therefore, such information could be used in order to prune large datasets that do not scale up.

Acknowledgments. The third author acknowledges the support of ANR grants ASPIQ (ANR-12-BS02-0003), QUALINCA (ANR-12-0012) and DURDUR (ANR-13-ALID-0002). The work of the third author has been carried out part of the research delegation at INRA MISTEA Montpellier and INRA IATE CEPIA Axe 5 Montpellier.

References

1. https://en.wikipedia.org/wiki/birthday_problem
2. http://wiki.dbpedia.org/downloads39
3. http://www.mpi-inf.mpg.de/yago-naga/yago/downloads.html
4. Atencia, M., Chein, M., Croitoru, M., David, J., Leclère, M., Pernelle, N., Saïs, F., Scharffe, F., Symeonidou, D.: Defining key semantics for the RDF datasets: experiments and evaluations. In: Hernandez, N., Jäschke, R., Croitoru, M. (eds.) ICCS 2014. LNCS, vol. 8577, pp. 65–78. Springer, Heidelberg (2014)
5. Atencia, M., David, J., Euzenat, J.; Data interlinking through robust linkkey extraction. In: ECAI 2014–21st European Conference on Artificial Intelligence, pp. 18–22 , Prague, Czech Republic - Including Prestigious Applications of Intelligent Systems (PAIS 2014), pp. 15–20, August 2014
6. Atencia, M., David, J., Scharffe, F.: Keys and pseudo-keys detection for web datasets cleansing and interlinking. In: ten Teije, A., Völker, J., Handschuh, S., Stuckenschmidt, H., d'Acquin, M., Nikolov, A., Aussenac-Gilles, N., Hernandez, N. (eds.) EKAW 2012. LNCS, vol. 7603, pp. 144–153. Springer, Heidelberg (2012)
7. Chen, P.Y., Popovitch, P.M.: Correlation: Parametric and Nonparametric Measures. Sage University Papers Series on Quantitative Applications in the Social Sciences (2002)
8. Husson, F., Lê, S., Pagé, J.: Analyse de données avec R, 2éme édition revue et augmentée (2016)
9. Gunopulos, D., Khardon, R., Mannila, H., Saluja, S., Toivonen, H., Sharma, R.S.: Discovering all most specific sentences. ACM Trans. Database Syst. **28**(2), 140–174 (2003)

10. Holmes, S.: Multivariate analysis: the french way, pp. 1–14 (2006)
11. Hyndman, R.J., Fan, Y.: Sample quantiles in statistical packages. Am. Stat. **50**, 361–365 (1996)
12. Pernelle, N., Saïs, F., Symeonidou, D.: An automatic key discovery approach for data. J. Web Sem. **23**, 16–30 (2013)
13. R Core Team: R: A Language and Environment for Statistical Computing. R Foundation for Statistical Computing, Vienna, Austria (2015)
14. Sismanis, Y., Brown, P., Haas, P.J., Reinwald, B.: Gordian: efficient and scalable discovery of composite keys. In: VLDB, pp. 691–702 (2006)
15. Soru, T., Marx, E., Ngomo, A.-C.N.: ROCKER - a refinement operator for key discovery. In: Proceedings of the 24th International Conference on World Wide Web, WWW 2015 (2015)
16. Symeonidou, D., Armant, V., Pernelle, N., Saïs, F.: SAKey: Scalable Almost Key discovery in RDF data. In: Mika, P., et al. (eds.) ISWC 2014, Part I. LNCS, vol. 8796, pp. 33–49. Springer, Heidelberg (2014)

A Model for Linked Open Data Acquisition and SPARQL Query Generation

Céline Alec[(✉)], Chantal Reynaud-Delaître, and Brigitte Safar

LRI, Univ. Paris-Sud, CNRS, Université Paris-Saclay, 91405 Orsay, France
{celine.alec,chantal.reynaud,brigitte.safar}@lri.fr

Abstract. Nowadays, Linked Open Data (LOD) represents a promising source for many applications of the Semantic Web. However, appropriate data acquisition techniques have to be developed to overcome both incompleteness and redundancy. Our work addresses this problem in the scenario of populating an OWL ontology with property assertions considering the existence of multiple, equivalent, multi-valued and missing properties in the LOD. Since the correspondences to be built are complex, we propose a model to specify them. We also define a model to specify alternative paths to access properties in case of missing values. We then show how these models are used to automatically generate SPARQL queries and thus, facilitate interrogation. A running example is given.

Keywords: Linked data acquisition · Complex correspondences · SPARQL query generation

1 Introduction

During the last years, new sources have emerged, like Linked Open Data (LOD), which are promising for many Semantic Web applications. However, LOD datasets reveal data quality problems such as incompleteness, redundancy, inconsistency [15]. Given these issues, specific acquisition techniques have to be developed. In this paper, we propose a model to support LOD data acquisition.

Our work is done in the settings of an ontology-based method [1] to discover definitions of concepts based on the individuals with known property assertions, using machine learning techniques. The quality of results depends on the inputs, particularly property assertions characterizing individuals. Having complete information is essential in order to achieve the best results as output of the entire process. First, property assertions are extracted from a corpus of texts. Then, as texts that are considered are incomplete for some required property assertions, LOD datasets are exploited. The approach is applicable to various domains. For a given area, the designer selects the most relevant datasets for the involved properties and indicates the properties to be valued. The number of these properties is assumed not to be too large, so this task can be manually done. However, the ontology and the LOD datasets differ in regards to vocabulary and structure. Complex correspondences have to be defined. In this work,

O. Haemmerlé et al. (Eds.): ICCS 2016, LNAI 9717, pp. 237–251, 2016.
DOI: 10.1007/978-3-319-40985-6_18

we propose a model to define them when datasets include redundant properties, missing values and multi-valuation. This is our first contribution.

This model is also a support to ease the access to LOD data. Automatic generation of SPARQL [6] queries can be conducted from this model. We assume that the structure of SPARQL queries can be determined by our acquisition model, in a domain independent way. This is our second contribution. Our work is innovative because complex processes needed to access information not explicitly represented in LOD datasets can be defined. Moreover, the querying part to get this information is completely transparent for the user. To our knowledge, those issues have not yet been addressed by the state-of-the-art.

Section 2 presents an example of complex data motivating our work. Section 3 reviews related works. The model of acquisition is presented in Sect. 4. Section 5 explains how it supports automatic generation of SPARQL queries, for which a running example is given in Sect. 6. We conclude in Sect. 7.

2   A Motivating Example

Let us suppose that we are looking for assertions of the property temperature in the ontology myOnto. This property represents the average temperature characterizing the concept Weather. Individuals of Weather correspond to meteorological data of a given place for a given month. Thus, myOnto:WeatherJanuaryNaples is an individual of Weather corresponding to the meteorological data of Naples in January. This individual is characterized by the two following assertions: <myOnto:Naples myOnto:hasWeather myOnto:WeatherJanuaryNaples>, <myOnto:WeatherJanuaryNaples myOnto:concernMonth myOnto:January>. In DBpedia [2], several properties, e.g., dbp:janHighC, dbp:janLowC, dbp:janMeanC, etc., can be used to establish a correspondence with temperature. Each property concerns a particular month and its value may be expressed with particular measure units (Celsius, Fahrenheit). Furthermore, each property may be multi-valued. For example, dbp:janMeanC has two values for the DBpedia resource Naples: 8.1 (xsd:double) and 8.7 (xsd:double). To find assertions of the property temperature for myOnto, a correspondence with DBpedia properties has to be established. An example of the correspondence could be between: (i) the source property temperature with the constraint that its domain concerns January (via the property concernMonth) and (ii) a target expression $Avg(Avg(janMeanC),\ Avg(Conv(janMeanF)),\ \frac{1}{2}(Avg(janHighC)+Avg(janLowC)),\ \frac{1}{2}(Avg(Conv(janHighF))+Avg(Conv(janLowF))))$.

Each Avg deals with multi-valuation by returning a unique value. This correspondence is complex because we have to deal with a source constraint (domain concerns January) and a target property resulting from many calculations and aggregations, therefore not explicitly represented in the dataset. Moreover, some places in DBpedia, like Alaska, have no valuation for any of the above properties. Nevertheless, alternative approximate values may be found, as the meteorological data about Juneau, the capital of Alaska. A close version of this example is used in Sect. 6 to explain the automatic query generation.

3 Related Work

Our work is related to several research fields, such as RDF data mediation, question answering and ontology alignment with complex matches.

Some proposals have been provided for adopting query reformulation to achieve data mediation in the Linked Data cloud and produce flexible mediators. [3] proposes a graph pattern rewriting algorithm that exploits graph rewriting rules and ontology alignments in order to translate SPARQL queries to run the same query on different data sets. It focuses on data manipulation functions to handle information represented and aggregated in different ways such as different unit measures or different representations of properties (i.e., address). These data manipulations that we also consider in our work, remain actually quite simple. A generic method for SPARQL query rewriting is provided in [9]. A set of predefined mappings between ontology schemas represented in OWL is exploited. Note that, in this work, a mapping containing aggregates was considered as meaningless since the current SPARQL version could not represent them. Our work is related since constraints are represented in OWL too, but we consider more complex correspondences and, above all, aggregates.

Query answering approaches have been proposed as intuitive ways of accessing RDF data [8]. Questions are translated into triples, which are matched against RDF data relying on some heuristics and similarity measures. Then triple patterns are executed as SPARQL queries. However, sometimes the translation process is not possible. The question answering system described in [14] focuses on complex situations with questions including specific determiners as more than or the most leading to HAVING or ORDER BY clauses in SPARQL. Their strength is to generate queries from SPARQL templates. We use this idea in our work too. In other proposals, such as [13], user queries are limited to keywords, which are mapped to IRIs. Then, graph patterns are generated and executed as SPARQL queries. Once again, we differ because, in addition, we aim at retrieving information that may not be explicitly represented. This implies that accessing first relevant data is needed to compute the required one. Relevant data may be useful in the calculation of the required data without being similar or equivalent to it. In that case, queries must often include sub-queries. Such complex queries can not be derived based on question analysis.

Finally, several ontology alignment approaches have emerged in the literature in the last years [4] and some of them involve complex correspondences. In [11,12], correspondence patterns are presented as tools facilitating definition of complex correspondences between ontologies. They can be used to derive complex correspondences from simple previously calculated ones. Very often, techniques proposed for complex matchings exploit simple matches generated beforehand. This is the case, for instance, of the approach proposed by [10] for complex datatype properties from two classes, relying on an instance-based matching technique to identify simple matches. Finally, [5] defines complex correspondences for rewriting query patterns. Indeed query rewriting requires complex 1:n, n:1 or n:m correspondences involving *equivalence*, *more general* and *more specific* relations and expressed using constructors of a formal language.

From these proposals, we see that complex correspondences are needed for query rewriting. The process to define them is complex, even if they are usually defined from simple ones. We differentiate ourselves on this point. A correspondence between a source entity and a target entity resulting from a complex process and involving several aggregates applied in succession, cannot be derived from a simple correspondence previously acquired. Furthermore, entities involved in such complex treatments cannot be discovered by any existing alignment system.

4   A Model for Linked Open Data Acquisition

We consider a source ontology O_s and a target ontology O_t. O_s is an OWL ontology. O_t is an ontology providing an access to RDF sources, such as LOD datasets. Each ontology is defined as a tuple ($\mathcal{C}$, $\mathcal{P}$, $\mathcal{I}$, $\mathcal{A}$) where $\mathcal{C}$ is a set of concepts, $\mathcal{P}$ a set of (datatype, object and annotation) properties characterizing concepts, $\mathcal{I}$ a set of individuals and property assertions, and $\mathcal{A}$ a set of axioms.

This section defines a model to acquire O_s property values from a LOD dataset. This task requires: (1) to find the individual i_t from $\mathcal{O}_t$ corresponding to the individual i_s from O_s to be characterized, (2) to find the properties from O_t, called target properties, corresponding to a property from $\mathcal{O}_s$ that needs to be valued, called source property. This paper is limited to the second point. It is assumed that a preliminary step resulted in a set $Ind_{s,t}$ of couples (i_s, i_t) where an individual i_s from the source ontology corresponds to an individual i_t from the target ontology. In the following, we present a model to acquire LOD property values composed of: (1) a model establishing correspondences between a source property and a target property, and (2) a model defining paths to access property values.

4.1   Model of Correspondences

The model of correspondences defines correspondences (PE_s, PE_t) between a source Property Expression (PE_s) and a target Property Expression (PE_t). A correspondence applies for each couple $(i_s, i_t) \in Ind_{s,t}$ such as: $i_s \in domain(PE_s)$ and $i_t \in domain(PE_t)$.

Definition 1. *A property expression in $\mathcal{O}_s$ (PE_s) is an object property (op) or a datatype property (dp), possibly with a domain restriction expressed in OWL-DL noted $PE_s.Restrict(d)$.*
$$PE_s = op|dp|PE_s.Restrict(d)$$

Example 1. The property temperature.(concernMonth VALUE January) is a PE_s. It concerns the datatype property temperature whose domain (Weather) is constrained by (concernMonth VALUE January) expressed in Manchester syntax [7]. This PE_s only characterizes i_s corresponding to January meteorological data.

Now, we define the notion of property expression in $\mathcal{O}_t$ (PE_t), starting with the notion of elementary property on which the definition of PE_t is based.

Definition 2. *An elementary property p_e is a property in $\mathcal{O}_t$ or its inverse.*
$p_e = op|dp|op^{-1}$

Definition 3. *A property expression in $\mathcal{O}_t$ (PE_t) is an elementary property (p_e) in $\mathcal{O}_t$, or an expression (f) using one or several property expressions in $\mathcal{O}_t$. A PE_t may include domain constraints ($PE_t.Constr(d)$), range constraints ($PE_t.Constr(r)$) or constraints on an element involved in a range constraint ($PE_t.Constr(f(Constr(r)))$).*
$PE_t = p_e|f(PE_t)|f(PE_t, PE_t)|PE_t.Constr(d)|PE_t.Constr(r)|PE_t.Constr$
$(f(Constr(r)))$

By recursion, a PE_t can be a function of n PE_t. For example, $PE_{t1} = f(PE_{t2}, f(PE_{t3}, PE_{t4})) = f(PE_{t2}, PE_{t3}, PE_{t4})$. The function f is specified by the designer. When f is unary, he has to indicate whether it is an operation of aggregation or an operation of transformation and to clarify its nature. For an operation of aggregation, he must choose between a counting operation, a sum, minimum, maximum, average, etc. For an operation that transforms values, the choice is between a mathematical calculation, a calculation of the length of a string, a change from uppercase to lowercase or vice versa, the calculation of the absolute value of a number, etc. If f is n-ary, the designer must indicate whether the function corresponds to a set-theoretic operation (union or difference) or an operation of transformation (mathematical calculation, concatenation, etc.).

Constr is any constraint expressible as a SPARQL graph pattern (possibly with a filter constraint). $Constr(f(Constr(r)))$ is a constraint on the aggregation (f) of a variable involved in a range constraint defined beforehand (cf. Example 3). Examples of target properties given in the following are extracted from DBpedia.

The valuation of a PE_t is defined as follows:
$val(PE_t) = val(p_e)|f(val(PE_t))|f(val(PE_t), val(PE_t))|val(PE_t.Constr(d))$
$|val(PE_t.Constr(r))|val(PE_t.Constr(f(Constr(r))))$

Example 2. Avg(janPrecipitationMm, janRainMm) is a PE_t whose value is the average of all values from the union of janPrecipitationMm and janRainMm. To make sure this average does not take into account negative values, a range constraint might be introduced to reject negative values, e.g., FILTER(r >=0).

Example 3. artist is a DBpedia property whose domain is MusicalWork and range is Agent. Thereby, artist^{-1} is a p_e (by extension a PE_t), which, combined with the range constraints r rdf:type dbo:Album and r dbo:genre ?gen, corresponds to album names (with a genre). Adding the constraint COUNT(DISTINCT ?gen)>=3 on the variable ?gen involved within the constraints mentioned above, requires that there are at least three different genres.

In summary, such a model is interesting because it specifies correspondences with PE_t that may have a very complex definition. However, this model alone is not enough to deal with missing values in the LOD datasets. The next section presents a path model to address this issue.

4.2   Path Model

Accessing Paths for a PE_t. The model of correspondences requires an access to values of PE_t. When a PE_t has a value, it can be directly accessed without any difficulties. When this is not the case, we propose an alternative solution because we are in a context where values for PE_t are essential, even if it is an approximation. Thus, when a PE_t of an individual i_t is not valued, we look for its value(s) for other individuals than i_t. For example, if we do not find the value of the mean temperature in Alaska, we can search it for Juneau, its capital. This value will be an approximation of what we are looking for. The access to this PE_t for i_t is not direct since another individual than i_t has to be reached first. This led us to define two types of access to a PE_t, a direct access and a composed access. We give in this section the definition of each access type. It is based on the notion of property path defined w.r.t. elementary property paths.

Definition 4. *An elementary n^{th}-order property path $(n \geq 1)$, $path_n^{elem}$, is a composition of object properties $(p^1.p^2 \ldots p^n)$. For a target individual i_t, it allows an access to a set of target individuals I_t in $\mathcal{O}_t$.*

The access to a set of target individuals using an elementary path is explained by properties composing the path that could be multivalued.

Example 4. In Fig. 1, country.capital is an elementary 2nd-order property path $(p^1.p^2)$. Applied to the individual Cephalonia in DBpedia, it accesses to the set $I_t = \{$Athens$\}$. capital is an elementary 1st-order property path (p^1). Applied to the individual Sri_Lanka in DBpedia, it accesses to the set $I_t = \{$Colombo; Sri_-Jayawardenepura_Kotte$\}$, because the property capital of Sri_Lanka is multivalued.

Definition 5. *An nth-order property path $(n \geq 1)$ is an elementary path or a set of elementary paths, possibly with a constraint on I_t and/or i_t.*
$path_n = path_n^{elem}|Set(path_n^{elem})|path_n.Constr(i_t)|path_n.Constr(I_t)$

Definition 6. *The access to a PE_t is **composed** if it is accessed by an nth-order property path $(n \geq 1)$, **direct** otherwise.*

Example 5. Let $PE_t = \text{Avg}(\text{janPrecipitationMm})$ correspond to the average of values of janPrecipitationMm. Its access is direct for Abu_Dhabi and composed for Sri_Lanka (cf. Fig. 1).

As said in Definition 5, a path can be a set of elementary paths. This is most likely to occur when redundant properties belong to the path. For example, *part*, $isPartOf^{-1}$ and *p* are three ways to express the notion of subpart in DBpedia. Moreover, constraints on i_t and/or I_t are possible. A path constraint may concern the initial individual or the set of individuals accessed by paths from it.

Example 6. Suppose we want to know the country of origin of a film. The PE_t will be the union of the corresponding properties in DBpedia, e.g., *country* and *nationality*. In case of missing values, the nationality of the director will be

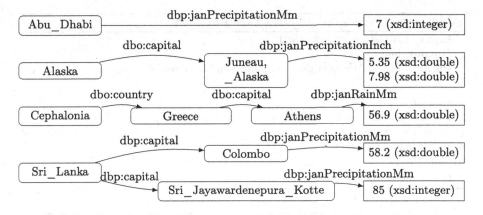

Fig. 1. Examples of paths to access to precipitation in DBpedia

a good approximation. Hence, Windy_City_Heat is a film without any country of origin in DBpedia. Nevertheless, knowing that Windy_City_Heat.director.nationality = Americans, its American origin can be deduced.

Constraint on i_t: If the designer believes that, with the globalization, many directors now realize films outside their own countries, he can add a constraint expressing that this approximation is valid only if the film is not recent (e.g., before 2000). This constraint is about the film (i_t), and can be expressed by the following graph pattern: i_t `dbp:released ?year. FILTER (?year <=2000)`.

Constraint on I_t: On the other hand, the designer may consider that these approximations will be quite good in general except for the films of some famous directors as James Cameron (Canadian) or Ridley Scott (British) who have made a lot of American films. Here, the constraint is related to the director (I_t) and can be expressed by the following graph pattern:
I_t `rdfs:label ?name. FILTER (LANGMATCHES(LANG(?name), "en")`
`&& !regex(?name, "Ridley Scott") && !regex(?name, "James Cameron"))`

Algorithms to Browse Access Paths. The designer defines all the access paths for each PE_t involved in correspondences with properties from $\mathcal{O}_s$, and assigns them a priority rating according to relevance. The Algorithm 1 implements this idea. The principle is to browse the paths leading to one PE_t by priority until a value of the PE_t is obtained. For that, the algorithm calls the function userAlgo composed of a switch part. An example of implementation of this switch part is presented after the function userAlgo. In the paths contained in this switch, `getVal()` is a function returning the PE_t values. For example, case 2 returns the PE_t values for the set of individuals being the capital of i_t, and *subparts* in cases 4 and 8 corresponds to the DBpedia properties isPartOf[1], part and p. Consequently, the path in case 4 corresponds to a set of three elementary paths. Finally, note that *?prop*, in these paths, means any DBpedia property.

Algorithm 1. Algorithm to extract data from $\mathcal{O}_t$

Input : i_t the target individual
Output: info to store in $\mathcal{O}_s$
Param : maxNbAttempts the maximum number of paths to browse

1 info $\leftarrow$ *null* ;
2 attempt $\leftarrow 1$;
3 **while** info $=$ *null and* attempt $\leq$ maxNbAttempts **do**
4 $\quad$ info $\leftarrow$ userAlgo(i_t, attempt) ;
5 $\quad$ attempt $\leftarrow$ attempt $+1$;
6 **end**
7 **return** info;

Function userAlgo

Input : i_t the target individual,
$\quad\quad\quad\quad$ attempt the number of attempt
Output: info to store in $\mathcal{O}_s$

1 info $\leftarrow$ *null* ;
2 **switch** *attempt* **do**
3 $\quad$ **case** i $\quad\quad\quad$ /* i=1 ...maxNbAttempts */ $\quad\quad\quad$ /* n^{th} order */
4 $\quad\quad$ info $\leftarrow i_t.p^1.p^2\ldots p^n.PE_t.\text{getVal}()$; break
5 **endsw**
6 **return** info;

Switch part: Example with data on Weather

---

1 **case** *1* info $\leftarrow i_t.PE_t.\text{getVal}()$; break; $\quad\quad\quad\quad$ /* direct access */
2 **case** *2* info $\leftarrow i_t.capital.PE_t.\text{getVal}()$; break; $\quad\quad\quad$ /* 1st-order */
3 **case** *3* info $\leftarrow i_t.largestCity.PE_t.\text{getVal}()$; break; $\quad\quad$ /* 1st-order */
4 **case** *4* info $\leftarrow i_t.subparts.PE_t.\text{getVal}()$; break; $\quad\quad\quad$ /* 1st-order */
5 **case** *5* info $\leftarrow i_t.country^{-1}.PE_t.\text{getVal}()$; break; $\quad\quad$ /* 1st-order */
6 **case** *6* info $\leftarrow i_t.?prop.PE_t.\text{getVal}() \cup i_t.?prop^-.PE_t.\text{getVal}()$; break;
$\quad\quad\quad\quad\quad\quad\quad\quad\quad\quad\quad\quad\quad\quad\quad\quad\quad\quad\quad\quad\quad\quad\quad\quad\quad\quad$ /* 1st-order */
7 **case** *7* info $\leftarrow i_t.country.capital.PE_t.\text{getVal}()$; break; $\quad\quad$ /* 2nd-order */
8 **case** *8* info $\leftarrow i_t.?prop^{-1}.subparts.PE_t.\text{getVal}()$; break; $\quad$ /* 2nd-order */

A same set of paths can be used for several PE_t. The parameter maxNbAttempts, fixed by the designer, indicates the maximum number of paths to browse for a particular PE_t. For instance, the example of the switch part presented is reusable to find values of latitude, longitude and monthly temperature and precipitation, but not necessarily with the same value of maxNbAttempts.

In our experiments on 80 destinations, 29 have a direct access to DBpedia weather data. Thanks to the switch part, every destination gets weather data.

5   Automatic Generation of Queries

The model for LOD acquisition presented in Sect. 4 supports the automatic generation of SPARQL 1.1 queries, particularly the generation of `CONSTRUCT` queries, returning an RDF graph (property assertions) that

```
CONSTRUCT { is ps ?val0 }
WHERE {it pathn.PEt ?val0}
```

we then add in $\mathcal{O}_s$. We note p_s the property of the individual i_s considered in PE_s and $?val_0$ the required values. The triples to be added in $\mathcal{O}_s$ are instances of the triple pattern $\{i_s\ p_s\ ?val_0\}$. The `WHERE` clause specifies how to get the value(s) of $?val_0$. This clause contains a graph pattern to be matched against the target graph, noted $\{i_t\ path_n.PE_t\ ?val_0\}$ such as (1) $(i_s, i_t) \in Ind_{s,t}$, (2) the PE_t corresponds to the PE_s according to the model of correspondences, (3) $path_n$ is a nth-order property path, possibly empty or defining a composed access to the PE_t.

The execution of queries follows the Algorithm 1 given in the previous section. For example, if the switch part contains 8 cases like the switch part example, 8 `CONSTRUCT` queries are created (1 per case) containing the variables i_s and i_t. Then, for each couple $(i_s, i_t) \in Ind_{s,t}$, we run the query of case 1. If its result is not empty, we continue with the next couple (i_s, i_t). Otherwise, we run the query of case 2 and so on until the result is not empty for the current couple.

We describe how to generate the content of the `WHERE` block of this query for a given PE_s, by adopting a pattern-based approach. All patterns generate parts of SPARQL expressions, which have to be inserted in the `WHERE` block. Note that if p_s is an object property, then a treatment, not mentioned here, is automatically done to transform the values obtained from the `WHERE` block into individuals in $\mathcal{O}_s$. First, we present the general process of automatic generation of the `WHERE` block. Second, we present the patterns used in this process.

5.1   Query Generation Process

We describe here how to generate the `WHERE` block of the `CONSTRUCT` query in SPARQL 1.1 aiming at adding property assertions in $\mathcal{O}_s$. It is a three-time process which takes as inputs definitions of correspondences and accesses supported by the model presented in Sect. 4 and two sets of patterns, one dealing with the **Format** of a PE_t and the other dealing with **Paths**.

The first step consists in generating the SPARQL part dealing with the format of the PE_t (by application of *Format patterns*). This step translates the expression "$i_t\ PE_t\ ?val_0$" into a SPARQL expression. The second step is a pre-treatment done before the insertion of the access path (search for the position of its statement, change of variable names). The last step inserts the access path (by application of *Path patterns*) "$i_t\ path_n.PE_t\ ?val$" into a SPARQL expression.

Note that variable names can change while applying a pattern, e.g., in $Pattern_{aggr}$. In the end, the variable returned by the final query remains $?val_0$.

5.2   Patterns

We present the patterns w.r.t. their category (*Format* and *Path*). The **Cond** part represents the condition for the application of the pattern, i.e., the context, and the **Action** part is the solution to be implemented.

Format Patterns. We have defined seven patterns based on the characteristics of the PE_t format (cf. Definition 3), called Format Patterns. They are used to generate the part of the query corresponding to "$i_t\,PE_t\,?val_0$". Their application depends on the definition of the PE_t. Hence, a pattern including several expressions "$i_t\,PE_t\,?val$" applies when the PE_t is defined using several PE_t.

Nesting patterns will be done in the same way than nesting PE_t, starting with the most external part. For example, if $PE_t = f(PE_t)$, the pattern considering the function f applies first. Then, its included expression "$i_t\,PE_t\,?val$" will be recursively replaced by application of the pattern considering the format specificities of PE_t, input of f. If the PE_t is constrained, the constraint pattern applies first. We list below the seven Format Patterns.

$Pattern_{elem}$

Cond: $PE_t = p_e$

Action: Replace p_e by the name of the elementary property.

$$\boxed{i_t\ p_e\ ?val_j.}$$

Note that if p_e is op^{-1}, then "$i_t\,p_e\,?val_j$" will be replaced by "$?val_j\,op\,i_t$".

$Pattern_{transfo_1}$

Cond: $PE_t = f(PE_t)$ with f a function of transformation.

Action: Replace f by its mathematical expression

$$\boxed{\begin{array}{l}\{i_t\ PE_{t_1}\ ?val_{j+1}\}\\ \texttt{BIND (f(}?val_{j+1}\texttt{) AS } ?val_j\texttt{)}\end{array}}$$

or the SPARQL function (STRLEN, LCASE, ABS, etc.) corresponding to the transformation function.

$Pattern_{aggr}$

Cond: $PE_t = f(PE_t)$ with f a function of aggregation.

Action: Replace f by the SPARQL 1.1 operator corresponding to the aggregation function:

$$\boxed{\begin{array}{l}\texttt{SELECT (f(}?val_{j+1}\texttt{) AS }?val_j\texttt{)}\\ \texttt{WHERE \{}\\ \quad i_t\ PE_t\ ?val_{j+1}\\ \texttt{\}}\end{array}}$$

COUNT, SUM, MIN, MAX, AVG, GROUP_CONCAT and SAMPLE.

$Pattern_{set}$

Cond: $PE_t = f(PE_t, PE_t)$ with f a set-theoretic operation.

Action: Replace f by the SPARQL operator UNION or MINUS depending on the case.

$$\boxed{\begin{array}{l}\{i_t\ PE_{t_1}\ ?val_j\}\\ \texttt{f}\\ \{i_t\ PE_{t_2}\ ?val_j\}\end{array}}$$

$Pattern_{transfo_2}$

Cond: $PE_t = f(PE_t, PE_t)$ with f an operation of transformation.

Action: Replace f by its mathematical

$$\boxed{\begin{array}{l}\{i_t\ PE_{t_1}\ ?val_{j+1_a}\}\\ \{i_t\ PE_{t_2}\ ?val_{j+1_b}\}\\ \texttt{BIND (f(}?val_{j+1_a},?val_{j+1_b}\texttt{) AS }?val_j\texttt{)}\end{array}}$$

expression or its SPARQL function (CONCAT, CONTAINS, etc.).

$Pattern_{constr}$

Cond: $PE_t = PE_t.Constr(d)|PE_t.Constr(r)$.

The constraint $Constr$ can be on the domain of the PE_t, i.e.,

$$\boxed{\begin{array}{l}\{i_t\ PE_t\ ?val_j\}\\ Constr\end{array}}$$

i_t, or on its range r, i.e., $?val_j$.

Action: Replace $Constr$ by its expression (graph pattern). If it is a range constraint, replace r by the variable name, i.e., $?val_j$.

$Pattern_{having}$

Cond: $PE_t = PE_t.Constr(f(Constr(r)))$

Action: Instantiate the expression from the HAVING clause w.r.t. the expressed constraint. Among other things, replace f by its SPARQL 1.1 operator : COUNT,

```
SELECT ?val_j
WHERE {
    i_t  PE_t  ?val_j
}
GROUP BY ?val_j
HAVING (f(?constr) operator value)
```

SUM, MIN, MAX, AVG, GROUP_CONCAT, SAMPLE. If the constraint variable ?constr, corresponding to $Constr(r)$, is in a nested query, we add this variable in each SELECT nesting it as well as in a GROUP BY clause.

Path Patterns. We have defined three patterns based on access paths (following the Definition 5). They apply to generate the expression "$i_t\, path_n\, ?ind_n$".

The application of these patterns follows a pre-processing step consisting in (i) selecting the position where the path expression must be declared and (ii) performing the appropriate changes of variable names. Indeed, we could directly replace all references to i_t by a path expression in the query resulting from applying Format Patterns. However, this path would be repeated many times in the query if the definition of PE_t includes lots of elementary properties. With the pretreatment step, the path is cited only once.

The position of the path depends on the form of the query under construction.

- If this one starts with a SELECT clause, the expression "$i_t\, path_n\, ?ind_n$" is inserted in the WHERE clause of this SELECT clause. This will be the case when the format of the PE_t is $f(PE_t)$ or $PE_t.Constr(f(Constr(r)))$, with f an aggregation function, eventually with domain and/or range constraints. We made this choice to simplify the readability of the query. This simplification will be helpful in future work. Indeed, we would like to add further treatments on our queries and the current WHERE block will be nested in other SPARQL expressions.

- If the query under construction does not start with a SELECT clause, "$i_t\, path_n\, ?ind_n$" is inserted at the beginning of the query.

In both cases, $?ind_n$, the variable corresponding to I_t, is added in all SELECT clauses and their result-set is grouped by this variable.

The second pretreatment consists in replacing i_t in the current version of the WHERE block by $?ind_n$.

The *Path Patterns* nest within each other following the definition of $path_n$. To instantiate the patterns, we start with the constraint pattern $Pattern_{path_{constr}}$ expressed in terms of $path_n$, which can nest every type of *Path Pattern*. The set pattern $Pattern_{path_{set}}$ expressed in terms of elementary paths $p^1.p^2 \ldots p^n$ can nest patterns of elementary path $Pattern_{path_{elem}}$.

$Pattern_{path_{elem}}$

Cond: $path_n = p^1.p^2 \ldots p^n$

Action: Replace properties

```
i_t  p^1  ?ind_1 .  ?ind_1  p^2  ?ind_2 .  \ldots ?ind_{n-1}  p^n  ?ind_n .
```

p^j by their name. Note that if p^j equals op^{-1}, then "$x\, p^j\, ?ind_j$" will be replaced by "$?ind_j\, op\, x$".

$Pattern_{path_{set}}$
Cond: $path_n = Set(p_1^1.p_1^2 \ldots p_1^n, p_2^1.p_2^2 \ldots p_2^n, \ldots, p_m^1.p_m^2 \ldots p_m^n)$

{i_t $p_1^1.p_1^2\ldots p_1^n$?ind_n} UNION {i_t $p_2^1.p_2^2\ldots p_2^n$?ind_n} UNION...UNION {i_t $p_m^1.p_m^2\ldots p_m^n$?ind_n}

Action: Adapt the number of operators UNION w.r.t the length of the set.

$Pattern_{path_{constr}}$
Cond: $path_n = path_n.Constr(i_t) \, vert \, path_n.Constr(I_t)$

i_t $path_n$?ind_n
$Constr$

Action: Replace $Constr$ by its expression. If the constraint is on I_t, I_t is replaced by its variable ?ind_n. The expression "i_t $path_n$?ind_n" is replaced by the application of the corresponding pattern w.r.t. the form of the path.

6 Running Example

The process to generate automatically SPARQL queries has been implemented. We present a running example. The property is close to the one given in Sect. 2. This running example serves to illustrate the contribution of our work compared to the state of the art. Indeed, to our knowledge, no work deals with complex expressions we have defined in this paper. [3,9] do not consider the aggregates introduced in version 1.1 of SPARQL. [9] believes that the work of aggregation can be seen as a post-processing task. This is true for simple cases, but if several aggregates apply, each one on the results of sub-queries, and if the results of these aggregations are then transformed, this is no longer true.

Let $PE_t = Avg(Avg(janMeanC), \frac{Avg(janHighC) + Avg(janLowC)}{2})$, corresponding to the mean temperature in January. The resulting value cannot be obtained by a post-treatment of a unique SPARQL query returning all the necessary data. Three independent queries are needed, each one returning values of one elementary property involved in the PE_t. Then, several manual steps have to be performed: (1) aggregating the values of each property, (2) calculating the results of the fraction, (3) aggregating the final values. Each step is based on the result of the previous one. Such a post-treatment is complex. It must be performed in stages and a great part is done manually. Our process is more convenient because it handles aggregates applied in succession and avoids the need to write complex SPARQL queries. Our running example is presented below.

The Fig. 2 shows the evolution of the construction of the WHERE block of the final CONSTRUCT query dealing with the expression "i_t $path_n.PE_t$?val_0". The path is composed of the set of the three elementary paths expressing subparts (1st-order). The left column (cf. ① to ⑤) expresses the application of *Format Patterns*. If the access is direct (no path), the end of this step is the end of the process. If the access is composed (using a path), the two following steps are performed: pretreatment (cf. ⑥) and application of *Path Patterns* (cf. ⑦ to ⑧). Each i_t has to be replaced with the name of a target individual, e.g., dbr:Naples.

① $Pattern_{aggr}$ is applied due to the format, $Avg(i_t \, PE_{t'} \, ?val_1)$, of the PE_t, and f is instantiated with AVG.

① $Pattern_{aggr}$

```
SELECT (AVG(?val1) AS ?val0)
WHERE { it PEt' ?val1 }
```

⇓

② $Pattern_{set}$

```
SELECT (AVG(?val1) AS ?val0)
WHERE {
  { it PEt1 ?val1 } UNION
  { it PEt2 ?val1 }
}
```

⇓

③ $Pattern_{elem}$ + $Pattern_{transfo2}$

```
SELECT (AVG(?val1) AS ?val0)
WHERE {
  { it dbp:janMeanC ?val1. } UNION {
    { it PEt2a ?val2a }
    { it PEt2b ?val2b }
    BIND ((?val2a+?val2b)/2 AS ?val1)
  }
}
```

⇓

④ $Pattern_{aggr}$ (2 times)

```
SELECT (AVG(?val1) AS ?val0)
WHERE {
  { it dbp:janMeanC ?val1. } UNION {
    {
      SELECT (AVG(?val3) AS ?val2a)
      WHERE { it PEt3a ?val3 }
    }
    {
      SELECT (AVG(?val3) AS ?val2b)
      WHERE { it PEt3b ?val3 }
    }
    BIND ((?val2a+?val2b)/2 AS ?val1)
  }
}
```

⇓

⑤ $Pattern_{elem}$ (2 times)

```
SELECT (AVG(?val1) AS ?val0)
WHERE {
  { it dbp:janMeanC ?val1. } UNION {
    {
      SELECT (AVG(?val3) AS ?val2a)
      WHERE {it dbp:janHighC ?val3.}
    }
    {
      SELECT (AVG(?val3) AS ?val2b)
      WHERE {it dbp:janLowC ?val3.}
    }
    BIND ((?val2a+?val2b)/2 AS ?val1)
  }
}
```

⇒
⑥ Pretreatment if composed access

```
SELECT ?ind1 (AVG(?val1) AS ?val0)
WHERE {
  {it dbo:part ?ind1} UNION
  {?ind1 dbo:isPartOf it} UNION
  {it dbo:p ?ind1}
  { ?ind1 dbp:janMeanC ?val1. } UNION {
    {
      SELECT ?ind1 (AVG(?val3) AS ?val2a)
      WHERE { ?ind1 dbp:janHighC ?val3. }
      GROUP BY ?ind1
    }
    {
      SELECT ?ind1 (AVG(?val3) AS ?val2b)
      WHERE { ?ind1 dbp:janLowC ?val3. }
      GROUP BY ?ind1
    }
    BIND ((?val2a+?val2b)/2 AS ?val1)
  }
}
GROUP BY ?ind1
```

⇑
⑧ $Pattern_{path_{elem}}$ (3 times)

```
SELECT ?ind1 (AVG(?val1) AS ?val0)
WHERE {
  {it p1^1 ?ind1} UNION
  {it p2^1 ?ind1} UNION
  {it p3^1 ?ind1}
  { ?ind1 dbp:janMeanC ?val1. } UNION {
    {
      SELECT ?ind1 (AVG(?val3) AS ?val2a)
      WHERE { ?ind1 dbp:janHighC ?val3. }
      GROUP BY ?ind1
    }
    {
      SELECT ?ind1 (AVG(?val3) AS ?val2b)
      WHERE { ?ind1 dbp:janLowC ?val3. }
      GROUP BY ?ind1
    }
    BIND ((?val2a+?val2b)/2 AS ?val1)
  }
}
GROUP BY ?ind1
```

⇑
⑦ $Pattern_{path_{set}}$

```
SELECT ?ind1 (AVG(?val1) AS ?val0)
WHERE {
  it path1 ?ind1
  { ?ind1 dbp:janMeanC ?val1. } UNION {
    {
      SELECT ?ind1 (AVG(?val3) AS ?val2a)
      WHERE { ?ind1 dbp:janHighC ?val3. }
      GROUP BY ?ind1
    }
    {
      SELECT ?ind1 (AVG(?val3) AS ?val2b)
      WHERE { ?ind1 dbp:janLowC ?val3. }
      GROUP BY ?ind1
    }
    BIND ((?val2a+?val2b)/2 AS ?val1)
  }
}
GROUP BY ?ind1
```

Fig. 2. Running example

② $Pattern_{set}$ is applied due to the format, $\text{Avg}(i_t\ PE_{t_1}\ ?val_1, i_t\ PE_{t_2}\ ?val_1)$, of the PE_t. The expression "$i_t\ PE_{t'}\ ?val_1$" is translated into the union of two PE_t named PE_{t_1} and PE_{t_2}.

③ Due to the format janMeanC of the PE_t, $Pattern_{elem}$ is applied. So, PE_{t_1} is instantiated with the property janMeanC. Due to the format of $(i_t\ PE_{t_{2_a}}\ ?val_{2_a} + i_t\ PE_{t_{2_b}}\ ?val_{2_b})/2)$, $Pattern_{tranfo2}$ is also applied. PE_{t_2} is then instantiated with a mathematical operation on two PE_t.

④ Due to the format $((\text{Avg}(i_t\ PE_{t_{3_a}}\ ?val_3) + \text{Avg}(i_t\ PE_{t_{3_b}}\ ?val_3))/2)$, $Pattern_{aggr}$ is applied twice. Each time, f is instantiated with AVG.

⑤ $Pattern_{elem}$ is applied for both $PE_{t_{3_a}}$ and $PE_{t_{3_b}}$ instantiated respectively with the properties janHighC and janLowC.

⑥ A pretreatment is done before the insertion of the path. The path expression "$i_t\ path_1\ ?ind_1$" is added in the WHERE clause. The variable $?ind_1$ representing I_t is added in all SELECT clauses (and GROUP BY clauses are added). Finally, all i_t from the former query are replaced by $?ind_1$.

⑦ $Pattern_{path_{set}}$ is applied, replacing the path expression by a set of three elementary paths.

⑧ $Pattern_{path_{elem}}$ is applied, instantiated with the three elementary paths: $part$, $isPartOf^{-1}$ and p. We get the final WHERE block of the CONSTRUCT query.

7   Conclusion and Future Work

We presented a work done in the settings of an ontology-based approach requiring population of property assertions. We proposed a method to extract property assertions by exploiting LOD datasets dealing with multiple and missing values. This method includes the definition of a model for Linked Open Data acquisition guiding in the definition of complex correspondences between source and target property expressions. This model also defines access paths which can be useful in case of missing values. Second, it includes a pattern-based approach to automatically generate SPARQL queries based on the acquisition model. This aspect is crucial because if definitions of target property expressions are complex, queries that must be written to access data will also be very complex. We implemented the query generation process and presented a running example.

Future works will consider knowledge from O_s, such as functionality or cardinality restrictions. Treatments, such as aggregation on top of the actual WHERE block or restriction of the number of values (LIMIT), have to be developed.

References

1. Alec, C., Reynaud-Delaître, C., Safar, B.: An ontology-driven approach for semantic annotation of documents with specific concepts. In: Sack, H., Blomqvist, E., d'Aquin, M., Ghidini, C., Ponzetto, S.P., Lange, C. (eds.) ESWC 2016. LNCS, vol. 9678, pp. 609–624. Springer, Heidelberg (2016). doi:10.1007/978-3-319-34129-3_37
2. Auer, S., Bizer, C., Kobilarov, G., Lehmann, J., Cyganiak, R., Ives, Z.G.: DBpedia: a nucleus for a web of open data. In: Aberer, K., et al. (eds.) ASWC 2007 and ISWC 2007. LNCS, vol. 4825, pp. 722–735. Springer, Heidelberg (2007)

3. Correndo, G., Salvadores, M., Millard, I., Glaser, H., Shadbolt, N.: SPARQL query rewriting for implementing data integration over linked data. In: EDBT/ICDT Workshops, pp. 4:1–4:11. ACM, New York (2010)
4. Euzenat, J., Shvaiko, P.: Ontology Matching. Springer, Heidelberg (2013)
5. Gillet, P., Trojahn, C., Haemmerlé, O., Pradel, C.: Complex correspondences for query patterns rewriting. In: Ontology Matching at ISWC (2013)
6. Harris, S., Seaborne, A., Prud'hommeaux, E.: SPARQL 1.1 query language. In: W3C Recommendation (2013)
7. Horridge, M., Drummond, N., Goodwin, J., Rector, A., Stevens, R., Wang, H.: The manchester OWL syntax. In: OWLED, vol. 216, Athens, Georgia, USA (2006)
8. Kaufmann, E., Bernstein, A.: Evaluating the usability of natural language query languages and interfaces to semantic web knowledge bases. Web Semant. **8**(4), 377–393 (2010)
9. Makris, K., Bikakis, N., Gioldasis, N., Christodoulakis, S.: SPARQL-RW: transparent query access over mapped RDF data sources. In: EDBT, pp. 610–613. ACM (2012)
10. Pereira Nunes, B., Mera, A., Casanova, M.A., Fetahu, B., P. Paes Leme, L.A., Dietze, S.: Complex matching of RDF datatype properties. In: Decker, H., Lhotská, L., Link, S., Basl, J., Tjoa, A.M. (eds.) DEXA 2013, Part I. LNCS, vol. 8055, pp. 195–208. Springer, Heidelberg (2013)
11. Scharffe, F.: Correspondence patterns representation. Ph.D. thesis, Innsbrück University (2009)
12. Scharffe, F., Zamazal, O., Fensel, D.: Ontology alignment design patterns. Knowl. Inf. Syst. **40**(1), 1–28 (2014)
13. Shekarpour, S., Auer, S., Ngomo, A.C., Gerbe, D., Hellmann, S., Stadler, C.: Keyword-Driven SPARQL Query Generation Leveraging Background Knowledge. In: WI-IAT. pp. 203–210. IEEE Computer Society, Washington, DC (2011)
14. Unger, C., Bühmann, L., Lehmann, J., Ngomo, A.C.N., Gerber, D., Cimiano, P.: Template-based question answering over RDF data. In: Hacid, M.S., Ras, Z.W., Zighed, D.A., Kodratoff, Y. (eds.) WWW, pp. 639–648. ACM (2012)
15. Zaveri, A., Maurino, A., Equille, L.B.: Web data quality: current state and new challenges. Int. J. Semant. Web Inf. Syst. **10**, 1–6 (2014)

Dealing with Incompatibilities
During a Knowledge Bases Fusion Process

Fabien Amarger[2], Jean-Pierre Chanet[1], Ollivier Haemmerlé[2],
Nathalie Hernandez[2], and Catherine Roussey[1(✉)]

[1] Irstea, UR TSCF Technologies et systèmes d'information pour les agrosystèmes,
9 Avenue Blaise Pascal, CS 20085, 63178 Aubière, France
{jean-pierre.chanet,catherine.roussey}@irstea.fr
[2] IRIT, UMR 5505, Université Toulouse – Jean Jaurès,
5 allées Antonio Machado, 31058 Toulouse Cedex, France
{fabien.amarger,ollivier.haemmerle,nathalie.hernandez}@univ-tlse2.fr

Abstract. More and more data sets are published on the linked open
data. Reusing these data is a challenging task as for a given domain, several data sets built for specific usage may exist. In this article we present
an approach for existing knowledge bases fusion by taking into account
incompatibilities that may appear in their representations. Equivalence
mappings established by an alignment tool are considered in order to
generate a subset of compatible candidates. The approach has been evaluated by domain experts on datasets dealing with agriculture.

Keywords: Knowledge acquisition · Knowledge base fusion · Incompatibilities

1 Introduction

We propose in this paper a new fusion process of several Knowledge Bases (KBs)
in order to extract consensual knowledge. We made the hypothesis that the final
KB has a better quality because it will contain more reliable knowledge than the
source KBs. Our fusion process starts by aligning KBs by means of an existing
alignment tool. Then we generate candidates which are sets of ontological elements from different aligned KBs. For each candidate, we evaluate its trust score.
Previous work [1] propose several trust measures, which evaluate consensus found
in existing KBs. Merging different KBs about the same domain may generate
errors since the KBs can propose different viewpoints on a same domain. In this
paper we focus on one kind of errors coming from complex mappings between
three or more ontological elements. When two KB are aligned, if two elements
are mapped, then it means that the two elements are judged equivalent with
a certain degree. When one element of the KB_A is aligned with N elements of
KB_B, we consider that N − 1 mappings are wrong and we propose a method to
select the reliable mapping from the N ones starting with the element of KB_A.
From a complex mapping, the set candidates that will result will be considered

O. Haemmerlé et al. (Eds.): ICCS 2016, LNAI 9717, pp. 252–260, 2016.
DOI: 10.1007/978-3-319-40985-6_19

as incompatible. We will therefore seek to discover the maximum of candidates subsets (extension) such that all candidates of an extension are compatible.

This paper is organized as follows: (1) generic presentation of our fusion process, (2) generation of set of compatible candidates called an extension and (3) evaluation with a use case on wheat taxonomy generation.

2   Overview of Our Fusion Process

Our fusion process is composed of four activities as shown in Fig. 1. It starts from several knowledge bases called *Source Knowledge Bases* (SKB). The *Mapping Generation*, based on an alignment tool, computes mappings between onto-logical elements belonging to distinct sources. The *Candidate Generation* builds candidates which are sets of ontological elements coming from different sources, considered as similar. The *Trust Computation* computes the trust scores of the candidates with respect to consensus and the reliability degree of the mappings. The *Discovery of the Optimal Extension* allows the generation of an extension representing the maximal subset of compatible candidates validated by an expert (optimal extension). The first three activities have been already presented in [1,2]. In this article, we focus on the discovery of the optimal extension.

Knowledge Base. We consider that a knowledge base KB is an oriented labelled multigraph. $V_{kb} = C \cup PropO \cup PropDT \cup I \cup L$ is the finite set of vertices belonging to the knowledge base kb dispatched into disjoint subsets such that C is the set of classes, $PropO$ is the set of object properties, $PropDT$ is the set of datatype properties, I is the set of individuals, L is the set of vertices representing literals, including labels.

Ontological Element. An ontological element oe is a vertex of a knowledge base which can be mapped by an automatic alignment tool. As far as we know, alignment tools can only map individuals or classes, but not yet properties or RDF triples. Thus we limit ontological elements to classes or individuals in our fusion process: $oe \in C \cup I$.

We also define a function $nature(oe)$ which returns the type of the ontological element, which can be "class" if $oe \in C$, "individual" if $oe \in I$, "null" otherwise.

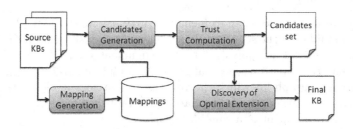

Fig. 1. Knowledge bases fusion process

254 F. Amarger et al.

Our work can be extended by considering as ontological elements other types of vertices. In that way, it could be possible to generate candidates in order to extract object properties or datatype properties.

Mappings. Assume SKB_i and SKB_j be two source knowledge bases used as input of our fusion process. We define a mapping m as an edge between a pair of vertices $\{oe_i, oe_j\}$, with $oe_i \in V_{skbi}, oe_j \in V_{skbj}$. A mapping represents an equivalence between two ontological elements, exhibited by means of the alignment tool. The mappings are defined in the following way:

- $V_{skbi} \neq V_{skbj}$: a mapping is always established between two vertices belonging to distincts SKB. $\nexists m = \{oe_i, oe_k\}$ such that $oe_i \in V_{skbi}$ and $oe_k \in V_{skbi}$.
- $nature(oe_i) = nature(oe_j) \neq$ "null". A mapping is always established between two ontological elements of same nature.
- $valueE : (C \cup I) \times (C \cup I) \rightarrow]0, 1]$ is a mapping which associates a unique reliability degree between 0 and 1 such that $valueE(oe_i, oe_j) = valueE(oe_j, oe_i)$ with each edge defined as a mapping.

In our work, we use the alignment tool LogMap[1] [6] since that system obtained good results during the evaluation OAEI 2014 [4] and, moreover, it allows to map individuals as well as classes.

The alignment systems allow to obtain mappings of type $1 : n$. In order to process these mappings, we made the hypothesis that such a mapping means that one element of a source is in relation of equivalence with one of the n elements of the other source, but the alignment tool could not make a choice and proposed a set of possible elements.

3 Generation of Compatible Candidates Set

On the basis of the hypothesis previously introduced, we first present the activity followed to generate candidates. Then, we explain how incompatibilities between candidates are identified. Finally we generate an extension which is a subset of compatible generated candidates.

3.1 Candidate Definition

A candidate *cand* represents an element that may belong to the final knowledge base. This candidate is the set of ontological elements extracted from the different SKBs and considered as equivalent by the alignment tool. We consider in the following that N SKBs have been aligned. A candidate $cand = (V_{cand}, E_{cand})$ is a non-oriented graph for which the vertices are ontological elements from the distinct SKBs and the edges are the mappings established by the alignment tool. We define the component of a candidate such as:

---

[1] http://www.cs.ox.ac.uk/isg/projects/LogMap/.

- $\forall v \in V_{cand}$ with $v \in V_{skbi}$ $\nexists v' \in V_{cand}$ such as $v' \in V_{skbi}$ et $v \neq v'$. All the vertices of a candidate belong to a different SKB. Therefore $|V_{cand}| \leq N$.
- E_{cand}: the edges of $cand$ are mappings.
- A candidate is a connected graph. $\forall v_1, v_2 \in V_{cand}$, there exists necessarily a chain $\{e_1, ..., e_k\}$ with $e_i \in E_{cand}$ linking v_1 to v_2. Therefore, all the vertices of $cand$ are linked to at least one other vertex of $cand$ by a mapping, which implies that all the vertices of $cand$ are of the same nature $\forall v_1, v_2 \in V_{cand}$ $nature(v_1) = nature(v_2)$.

Figure 2 presents two candidates for which elements are of the nature "individual". The two candidates – "Triticum" and "Triticum Durum" – represent ontological elements that can potentially belong to the final knowledge base. Edges drawn with dashes represent mappings with their reliability degrees.

3.2 Candidate Generation

From the set of sources and their alignments (such as presented in the background of Fig. 2) is extracted a multi-partite non-oriented graph of which the vertices are ontological elements of the different sources and edges are the mappings. All the connected components of this graph are computed. We consider that the minimal condition for identifying a consensus is if the candidate contains at least two ontological elements belonging to two distinct sources.

3.3 Incompatibilities

As explained previously, our process looks for connected components. This algorithm does not exclude the fact that an ontological element belongs to several candidates. This results from the fact that alignment tools identify $1 : n$ mappings, which means that an ontological element is mapped to 1 or several elements of the second knowledge base. We consider that these mappings are

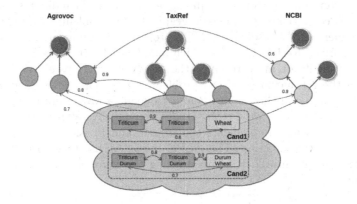

Fig. 2. Example of two candidates extracted from three sources

equivalence relations that the alignment tool wrongly established apart from one. We then define an incompatibility when several candidates share a common ontological element as shown in Fig. 3. We define an incompatibility as the couple of candidates that satisfy the following constraint: $Inc = \{Cand_1, Cand_2\}$ such that $\exists oe_{com}, oe_{com} \in V_{Cand_1}, oe_{com} \in V_{Cand_2}$.

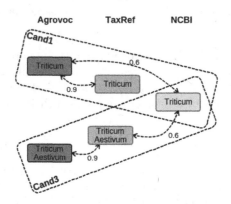

Fig. 3. Example of incompatibility between two candidates

Then we can define the non-oriented graph that represents the incompatibilities $G_{inc} = \{V_{inc}, E_{inc}\}$ such as the graph vertices V_{inc} are candidates and the graph edges E_{inc} represent an incompatibility between the connected vertices.

3.4   Extension Generation

By considering the incompatibility graph, we generate its complementary graph (the compatibility graph) that links only candidates that are not incompatible. Then we aim at generating the maximum cliques of the compatibility graph to obtain the extensions. Each maximum clique is a subset of candidates, each compatible. This corresponds to a classic graph theory problem: MCE (Maximum Clique Enumeration). To solve this MCE problem, several algorithms exist. The most used is Bron Kerbosch [3] for which many enrichment have been proposed such as Tomita [9] or more recently the algorithm of Eppstein et Strash [5]. Because of the NP-hardness of the problem, we do not aim at generating all the possible maximum cliques: our goal is to define a way of obtaining the optimal extension. To do so, we use the CSP solver GLPK (https://www.gnu.org/software/glpk/), which implements the Branch and Bound algorithm by maximising an objective function. The GLPK model we use tries to maximize the number of elements in *ext*, thus maximizing the number of candidates in the extension.

3.5 Finding the Optimal Extension

Thanks to our constraint model, we can obtain an extension among the possible ones. Our goal is to obtain the optimal extension that is the more likely to correspond to the expected knowledge base. To do so, we propose the expert to validate the generated extension. Then the process takes his/her opinion iteratively into account in order to add new constraints to the model and converge on an optimal solution for which all the candidates are correct.

During the validation step, all the candidates belonging to the current generated extension are presented to the expert. If the expert validates a candidate, a new constraint stating that the optimal extension must contain this candidate is added. For all non-validated candidates, the algorithm is run again in order to find a new extension. The extension is considered as optimal if all the candidates have been evaluated.

With this process, we can present to the expert a minimal number of candidates to evaluate. We deduce automatically that all incompatible candidates with a validated candidate must not belong to the optimal extension. The number of interactions with the expert is thus reduced. The time needed for the evaluation by the expert (number of interaction) is in the worst case equal to the number of generated candidates. When an incompatibility occurs, this number decreases.

4 Experiment About Wheat Taxonomy

Our evaluation use case is about the creation of a knowledge base on cereals. More information about this project are available in [1,8]. In this evaluation we only take care of the creation of a wheat taxonomy.

Our experts have chosen the following sources: (i) *Agrovoc*, the multilingual thesaurus managed by the FAO[2]. It contains over 40.000 terms, (ii) *TaxRef*, the french national taxonomic reference about living species managed by the national natural history museum[3]. It contains over 80.000 taxa, (iii) *NCBI Taxonomy*, the taxonomy created by the National Center for Biotechnology Information of United States[4]. It contains 1 million of taxa.

These sources have been transformed in knowledge bases using our SKOS transformation method [1], based on ontological module to drive the transformation process and define transformation patterns dedicated to each source. For our experiment, we use the ontological module called AgronomicTaxon[5] that merges several ontology design patterns [8]. Based on these three new knowledge bases, we can apply our activity of candidate generation. Then, we generate the graph of incompatibility. Table 1 presents data about this graph.

To evaluate our method, an expert evaluated candidates and we counted the number of interactions (validation or invalidation) that were required to

---

[2] http://aims.fao.org/standards/agrovoc/about.

[3] http://inpn.mnhn.fr/programme/referentiel-taxonomique-taxref.

[4] http://www.ncbi.nlm.nih.gov/.

[5] https://sites.google.com/site/agriontology/home/irstea/agronomictaxon.

Table 1. information about graph of incompatibility

Sources	$|eos|$	$|cands|$	$|incomps|$
Agrovoc	11	150	1555
TaxRef	19		
NCBI	130		

eos: ontological elements.

Table 2. Results

$|ext_{max}|$	$|ext_{opti}|$	Nb interactions	Ratio
25	23	62	0.41

ext_{max}: nb of candidates in the largest extension,

ext_{opti}: nb of candidates in the optimal extension,

$Ratio = \frac{nb\ interactions}{|cands|}$.

obtain the optimal extension. Table 2 presents these results. The size of the largest extension generated by our algorithm without manual evaluation was 25 candidates. The size of the extension manually evaluated was 23 candidates. Thus, 2 sets of incompatible candidates were not validated by the expert. None of the candidates pairwise incompatible were validated by the expert. That means that the $1 : n$ mappings between ontological elements were wrong. The expert needed 62 interactions to build the final extension. That means that he had to validate or invalidate 62 candidates among the 150 possible candidates. The ratio is 0.41. 41% of candidates had to be observed by expert to build the optimal extension. Thus our method divided by 2 the number of candidates to build the optimal extension.

Then we calculated the coverage and conciseness metrics defined in [7]. The redundancy metric is not relevant for our fusion process because we know that we have eliminated redundancy. This is the main advantage of our fusion process.

The coverage is the ratio between the number of the candidates in the extension (23) and the number of ontological elements extracted for each source. We obtained **Agrovoc:** 2.09, **TaxRef:** 1.12 and **NCBI:** 0.17.

The conciseness evaluates if the extension is the minimal representation of knowledge. This metric is useful to detect fusion process that merge data without aggregation. Conciseness calculates the ratio between the size of the extension (23) and the sum of elements extracted from sources (160). We obtained the value 0.14. Thus, we can conclude that the optimal extension is close to the minimum representation of knowledge.

During these evaluations, a phenomenon appeared that could significantly reduce the number of expert interactions. Indeed, several incompatible candidates are successively presented to the expert until he validates one. All these incompatible candidates come from the same complex mapping $1 : n$. We notice that the expert did not validate the candidates which contained elements that had less labels in common. Furthermore he validated the candidate that contains elements which had more labels in common. It would be interesting to present first the candidate that contains elements which have more labels in common. This idea could be generalized by taking into account not only the labels but the neighborhood. We can add these information in the objective function. Thus we would first present the candidates that contain elements which have the more neighbours (vertices or edges) in common.

5   Conclusion and Future Works

This paper presents a process to manage incoherencies between knowledge bases in a fusion process. The fusion process generates candidates that group similar elements extracted from distinct knowledge bases. We define what are incompatible candidates. These incompatibilities help to identify subsets of compatible candidates, which are called extensions. A method to find the optimal extension is proposed. Incompatibilities between candidates are also used during manual evaluation in order to limit the number of candidates that should be evaluated by experts. These methods have been validated on a real use case. The use case creates a knowledge base on plant taxonomy from multiple sources.

Our method of extension generation can be improved by taking into account the trust score of candidates. Our previous works present several functions to compute the trust score [1]. These scores can be used to optimize the objective function of our extension generation algorithm. Another interesting perspective should be to work on edge candidates. An edge candidate is composed of two candidates linked by edges having the same label (the same property). Our extension generation method should integrate neighboring candidates and edge candidates in order to faster the optimal extension discovery.

References

1. Amarger, F., Chanet, J.-P., Haemmerlé, O., Hernandez, N., Roussey, C.: SKOS sources transformations for ontology engineering: agronomical taxonomy use case. In: Closs, S., Studer, R., Garoufallou, E., Sicilia, M.-A. (eds.) MTSR 2014. CCIS, vol. 478, pp. 314–328. Springer, Heidelberg (2014)
2. Amarger, F., Chanet, J.-P., Haemmerlé, O., Hernandez, N., Roussey, C.: Construction d'une ontologie par transformation de systèmes d'organisation des connaissances et évaluation de la confiance. Ingénierie des Systèmes d'Information **20**(3), 37–61 (2015)
3. Bron, C., Kerbosch, J.: Algorithm 457: finding all cliques of an undirected graph. Commun. ACM **16**, 575–577 (1973)
4. Dragisic, Z., Eckert, K., Euzenat, J., Faria, D., Ferrara, A., Granada, R., Ivanova, V., Jiménez-Ruiz, E., Kempf, A.O., Lambrix, P., et al.: Results of the ontology alignment evaluation initiative 2014 (2014)
5. Eppstein, D., Strash, D.: Listing all maximal cliques in large sparse real-world graphs. In: Pardalos, P.M., Rebennack, S. (eds.) SEA 2011. LNCS, vol. 6630, pp. 364–375. Springer, Heidelberg (2011)
6. Jiménez-Ruiz, E., Cuenca Grau, B.: LogMap: logic-based and scalable ontology matching. In: Aroyo, L., Welty, C., Alani, H., Taylor, J., Bernstein, A., Kagal, L., Noy, N., Blomqvist, E. (eds.) ISWC 2011, Part I. LNCS, vol. 7031, pp. 273–288. Springer, Heidelberg (2011)
7. Raunich, S., Rahm, E.: Towards a benchmark for ontology merging. In: Herrero, P., Panetto, H., Meersman, R., Dillon, T. (eds.) OTM-WS 2012. LNCS, vol. 7567, pp. 124–133. Springer, Heidelberg (2012)

260 F. Amarger et al.

8. Roussey, C., Chanet, J.P., Cellier, V., Amarger, F.: Agronomic taxon. In: WOD, p. 5 (2013)
9. Tomita, E., Tanaka, A., Takahashi, H.: The worst-case time complexity for generating all maximal cliques and computational experiments. Theoret. Comput. Sci. **363**, 28–42 (2006)

Author Index

Printed in the United States
By Bookmasters